GEORGES CUVIER, FOSSIL BONES, AND

GEOLOGICAL CATASTROPHES

GEORGES CUVIER,

FOSSIL BONES,

and GEOLOGICAL

CATASTROPHES

———

New Translations & Interpretations

of the Primary Texts

———

MARTIN J. S. RUDWICK

The University of Chicago Press ∾ Chicago and London

The University of Chicago Press, Chicago 60637
The University of Chicago Press, Ltd., London
© 1997 by The University of Chicago
Published 1997
Printed in the United States of America

06 05 04 03 02 01 00 99 98 97   1 2 3 4 5

ISBN-13: 978-0-226-73106-3 (cloth)
ISBN-13: 978-0-226-73107-0 (paper)
ISBN-13: 978-0-226-73108-7 (e-book)
DOI: https://doi.org/10.7208/chicago/9780226731087.001.0001

Library of Congress Cataloging-in-Publication Data

Rudwick, M.J.S.
    Georges Cuvier, fossil bones, and geological catastrophes: new translations
and interpretations of the primary texts / Martin J. S. Rudwick
        p.   cm.
    Includes bibliographical references and index.
    ISBN 0-226-73106-5 (acid-free paper)
    1. Mammals, Fossil.   2. Catastrophes (Geology)   3. Geology—
History—18th century.   I. Cuvier, Georges, baron, 1769–1832.
Essays. English. Selections. II. Title.
    QE881.R83   1997
    569—dc21   97-15628
                                                              CIP

⊚ This paper meets the requirements of ANSI/NISO Z39.48-1992
(Permanence of Paper).

In memory of William Coleman

(1934 – 1988)

CONTENTS

Geological catastrophes are back in fashion. Fossil bones have rarely been out of fashion, particularly if they belong to dinosaurs. But only in the past twenty years have earth scientists felt able to explore the possibilities of linking the two, without fear of being dismissed as mavericks or cranks. In doing so, they are—whether they know it or not— reviving ideas that were first made the center of scientific debate by Georges Cuvier, just two hundred years ago.

Georges Cuvier (1769–1832) was by any reckoning a towering figure in early nineteenth-century science. Although he was primarily a comparative anatomist, and one of outstanding importance in the history of biology, his pioneer research on *fossil* mammals led him into what was then the self-consciously new science of geology. He argued strongly for the reality of extinction, and he linked this with a view of geological change that stressed the effects of occasional sudden physical events or "catastrophes" at the earth's surface. He was not the first to propound this kind of "catastrophism" (as it was later termed), but his arguments gave it powerful support and continued to be influential long after his death. More generally, however, Cuvier saw his research as "bursting the limits of time," by making it possible to reconstruct a reliable and detailed *history* of the earth and its life, back beyond the most recent "catastrophe" and long before the beginnings of human records or even the existence of human beings. This "geohistorical" perspective (as it would now be called) was arguably an even more important legacy to science than his catastrophism.

Cuvier's published papers made these themes widely known among those who would now be called scientists, but his work had a far wider impact than that. In particular, the eloquent "Preliminary discourse" that he prefixed to his great four-volume *Recherches sur les ossemens fossiles* (Researches on fossil bones, 1812) was easily accessible to any educated person, and it was immensely influential in the intellectual life of the Western world for the rest of the century. It was reissued as a short book, repeatedly reprinted, and translated into all the main scientific languages of the day. In addition to its arguments for geological "catastrophes" in the distant past, its treatment of extinction and adamant rejection of "transformist" explanations of the origin of species were vital components of the evolutionary debates that continued throughout the century.

Historical understanding and appreciation of Cuvier's work was until recently stultified by the perception that he had been doubly on the wrong side: wrong in his opposition to organic evolution and wrong in his claims for the reality of catastrophes. But modern approaches to the history of science, reinforced by the renewed acceptability of catastrophism among modern scientists, have now begun to restore Cuvier to his proper and prominent place in the history of science. It has been difficult, however, for nonspecialists to understand or appreciate the huge impact of his work, in the absence of an accessible edition of the relevant texts. This volume is designed to fill that striking gap in the literature available to English-speaking geologists and paleontologists, historians of science and intellectual historians, and students in those fields. It offers in translation not only the first edition of the "Preliminary discourse" itself, but also a selection of earlier articles and lectures; and these texts are accompanied by reproductions of some of the original illustrations.

This historical material is set in context by an editorial narrative and commentary. My aim has been to provide just enough biographical background and explanation of the texts, to allow Cuvier's words and images to speak for themselves. Specifically, I have tried to bear in mind the very different background knowledge of, say, geologists and paleontologists on the one hand, and historians of science and nineteenth-century culture on the other. This book is not a biography of Cuvier, nor does it even claim to be a definitive account of his work on fossils; so I have not thought it appropriate to document every point with references to the specialist historical literature. The most important sources, however, are described briefly in "Further Reading." I have also identified the publications that Cuvier cited—often in cryptic, abbreviated form—in his texts and footnotes; they are assembled here in a bibliography in modern style, to facilitate further reference.

The texts have been selected primarily to illustrate Cuvier's *geological*, not his zoological, work. But in his publications there was much overlap between those two fields, for reasons that will be clear in the texts themselves. Zoologists and historians of biology will therefore find here many of Cuvier's most important theoretical statements about the relation between form and function in the living organism, about the reality and significance of extinction, and about the implausibility of the kind of evolutionary theory that was being propounded during his lifetime. (If this book is not primarily about Cuvier's biology, still less is it concerned, except indirectly and in passing, with his work on the human sciences; it has no bearing on arguments over his position in relation to such modern concepts as racism and sexism.)

As far as I am aware, most of the shorter texts printed in this volume have never previously been available in English, and the few that appeared in British periodicals in Cuvier's time were poorly translated. I have made new translations of them all (two from German, all the others from French). Two brief but important texts have never been published even in their original French; in these cases the manuscripts (now in archives in Paris) are transcribed in an appendix.

By contrast, the "Preliminary discourse" has been well known in the English-speaking world through the translation that was commissioned by Robert Jameson, the Edinburgh professor whose boring manner later gave the young Charles Darwin a temporary aversion to geology. Jameson's edition of Cuvier's "Preliminary discourse" was first published in 1813, the year after the French original, and it was repeatedly updated and enlarged; it has been reprinted in the twentieth century, and has been used by many of the English-speaking historians who have written about Cuvier's work. That its style and diction are archaic would not matter much, but unfortunately the translation is often misleading and in places downright bad. I have therefore made a completely new translation into modern English. Even more seriously, however, Jameson's prefaces and his extensive editorial notes have been responsible for major distortions in the understanding of Cuvier's ideas, at least in the anglophone world. In particular, it was Jameson's comments, rather than Cuvier's own text, that led to the widespread belief that Cuvier had constructed his theories in order to support a literalistic interpretation of Genesis or to bolster the historicity of the biblical story of the Flood. I hope my translation will help to correct that gross misconception.

The preparation of this edition has taken me back many years to my first encounter with Cuvier's work on fossils. At that time I was a paleontologist, not a historian, and my initial interest in Cuvier was strictly scientific. E. S. Russell's great classic work *Form and function* (1916) had given me the clue that Cuvier's work, despite being antievolutionary and pre-Darwinian, might contain ideas that I could usefully apply in my own research on the functional evolution of some fossil invertebrates (the brachiopods). That hope was fulfilled beyond all my expectations: it turned out to be well worthwhile for an evolutionary paleontologist to become—at least for a time—imaginatively a pre-Darwinian.

My excursion into history had an unforeseen consequence. Cuvier's superb engravings of fossil bones, and the dusty, musty tomes in which I read his work, began to be as alluring to me as the beautiful sculptural shapes of fossil brachiopods. At the same time, and not by coincidence, the intellectual challenges of the history of science began to be as compelling as those of paleontology. Eventually, several years later, practical constraints forced me to choose between those two equally fascinating fields of research. I still feel a soft spot for Cuvier, as the historical figure who nudged me toward that point of decision.

Of all the friends and colleagues to whom I am grateful for help with this project, William Coleman was literally first and foremost. He generously shared with me his unrivalled knowledge of Cuvier's work, while he was writing his pioneering book *Georges Cuvier zoologist* (1964). His infectious enthusiasm for the history of science, and particularly for Cuvier studies (though he told me he had grown to dislike Cuvier the man), helped make me decide to become a historian of science too. I dedicate this book to his memory, as an inadequate token of what I owe him.

Other Cuvier scholars—and, equally, those primarily interested in his great opponent Lamarck or more generally in the history of geology and paleontology—have helped and encouraged me in ways too various to specify here. Among them are Michael Benton (Bristol), Chip Burkhardt (Urbana), Claudine Cohen (Paris), Pietro Corsi (Florence), François Ellenberger (Paris), Gabriel Gohau (Paris), Steve Gould (Cambridge, Mass.), Goulven Laurent (Angers), Dorinda Outram (Cambridge), Rhoda Rappaport (Poughkeepsie), Jim Secord (Cambridge), Ken Taylor (Norman, Okla.), and Bert Theunissen (Utrecht). Paul Pickowicz (San Diego) kindly helped me with Chinese history, and Noel Swerdlow (Chicago) with astronomy.

Charis Cussins (Ithaca, N.Y.) has worked closely with me on most of the translations. She has suggested literally hundreds of possible improvements—both stylistic and semantic—to my draft translations from the

French; we have discussed them in detail, and the texts reflect her invaluable contributions, but of course I take full responsibility for the final choice of words.

This book is part of a larger project to explore the emergence of a sense of the *history* of the earth—and specifically of a long and complex *prehuman* geohistory—in the late eighteenth and early nineteenth centuries. *The great Devonian controversy* (1985) traces the course of one particularly knotty problem in great technical detail; it is intended in part to show the kind of expert argument that lay behind the construction of every part of geohistory. Conversely, *Scenes from deep time* (1992) explores the origins of the new pictorial genre—of reconstructed "prehistoric monsters" in their environments—which first made that sense of geohistory vividly real to the general public. The present book focuses on the work of the person who, in my opinion, was one of the two most important figures in the larger story (a series of earlier articles dealt with the other, Charles Lyell). A future volume will, I hope, place both Cuvier's and Lyell's work in its context, by reconstituting the research practices that enabled them and many others to work out how to construct a history of the earth and its life on reliable foundations.

This long-term project has been generously supported for many years by grants from the National Science Foundation (SES-8705907/8896206, DIR-9021695, and SBR-9319955) and from the Academic Senate of the University of California, San Diego. While I was completing the present book in 1994–95, my work was also supported by a fellowship from the John Simon Guggenheim Memorial Foundation, a University of California President's Fellowship in the Humanities, and a visiting fellowship at Clare Hall, University of Cambridge.

I am grateful to the librarians and staff at the libraries of the Institut de France and the Muséum National d'Histoire Naturelle in Paris, for giving me access to their rich Cuvier archives, and for permission to print transcriptions of two previously unpublished manuscripts. The latter library has also generously given permission for the reproduction of two of Cuvier's manuscript drawings (figs. 8, 11). Other illustrations are reproduced from printed sources by kind permission of the Syndics of Cambridge University Library (figs. 4–7, 9, 10, 12, 14, 16–19, 23, 24); the remainder are by the author.

Finally, I am, as always, grateful beyond measure to Susan Abrams for her support and encouragement: her editorial skill and judgment, linked to the outstanding work of her colleagues, have rightly given the University of Chicago Press an unparalleled reputation among historians of science.

NOTES ON THE TEXTS

Translation is ultimately an impossible art, and scientific texts are hardly less impossible to translate satisfactorily than literary ones. I have aimed at translations that are as close as can be to what I believe to have been Cuvier's *meaning*, in the light of his other work and that of his contemporaries. (I do not subscribe to the current fashions of postmodernism, at least in the cop-out forms that deny the accessibility, relevance, or even existence of authorial intentions.) Where Cuvier's meaning was difficult to express in simple English—the translator's perennial dilemma—my compromises have leaned toward clarity of meaning rather than elegance of style. I have not hesitated to split his often overlong sentences, to run together his often overshort paragraphs, and to make other such adjustments that I judged would improve the readability of the whole. A few editorial additions of words that clarify the meaning are enclosed in brackets.

I have given particular care to the translation of Cuvier's key terms and phrases, and above all to those that embody his key metaphors; in such cases his own words are recorded, also in brackets. Readers who want to check the translations in greater detail can of course consult the original texts: several are available in modern reprints (see "Further Reading"), most of the others can be found in libraries that hold early nineteenth-century periodicals, and the two that have not been published previously are printed in an appendix in this volume.

I have thought it inappropriate in this historical work to try to give the correct *modern* names of all the fossil genera and species mentioned by Cuvier; to do so would in many cases have involved a quite different kind of historical research, on his original specimens. Readers who are paleontologists should have little difficulty in recognizing, at least approximately, what kinds of fossil Cuvier had in mind. Likewise I have followed the custom of Cuvier (and his contemporaries) in often citing the names of fossil animals informally, without italics or initial capitals.

References to works cited by Cuvier, and others mentioned in the editorial sections, are given in the footnotes in abbreviated form; the "Bibliography of Cuvier's Sources" lists them in full. Footnotes *to the texts* are by Cuvier, except for editorial material enclosed in brackets.

Readers familiar with French will notice that Cuvier, like his contemporaries, used some now obsolete forms of spelling (e.g. the "ossemens" in the title of his most famous work); I have of course followed his spelling scrupulously in my transcriptions of unpublished texts (appendix) and in the French words that I note in my translations.

I

THE THEORY OF THE EARTH

———————

Georges Cuvier was born in 1769 in the town of Montbéliard, which was at that time the center of a small French-speaking Protestant territory belonging to the duchy of Württemberg.[1] This in turn was one of the many separate states that were united in the following century to form what is now Germany. So when, as a young man, Cuvier arrived in Paris to make a career for himself in the sciences, he was doubly an outsider. He was not a Frenchman by birth, though he had found himself becoming one when Montbéliard was annexed by France during the Revolution; and his cultural affinities were with the small Protestant minority in France, rather than with the dominant Catholic culture that most of his Parisian colleagues—even if they were strongly anticlerical—had in their bones.

On the other hand, his origins gave him one great advantage. His modest bourgeois family, and particularly his mother, had the ambition and respect for education that were common in that social class, and

---

1. He was baptized Jean-Léopold-Nicholas-Frédéric, and much later added Dagobert; but after his elder brother Georges Charles Henri died in early childhood he adopted the name Georges, and used it—usually on its own—throughout his life. It is not difficult to imagine how that confused identity could have contributed to his unquestionably complex personality. The section "Further Reading" describes the main historical works on which the biographical outlines in this book have been based.

FIGURE I A portrait sketch of the young Cuvier, possibly drawn when he left home to become a student in Stuttgart. This lithograph was made—probably much later—from a drawing by Cuvier's uncle, the municipal architect in Montbéliard.

soon recognized Cuvier's exceptional talents. As a teenager he gained a place at Württemberg's main institution of higher education, the Karlsschule in Stuttgart (fig. 1). There he received a rigorous training in a broad range of subjects that were regarded as useful for the state's future civil servants. Some scientific subjects were included, and these added to Cuvier's already active interests in natural history. Above all, however, he had to become fluent in German, which made a second major European culture accessible to him. It was an advantage shared by few of his future French colleagues, most of whom were complacent in their command of what was then the world's premier scientific language, just as many English-speaking scientists are today.

When Cuvier graduated, no official position was immediately available for him in Württemberg, yet he needed to support his parents. Like some other young intellectuals in that situation, he therefore took employment as a private tutor in an aristocratic family. In Cuvier's case the family was that of the duke d'Hericy, a Protestant nobleman with estates in Normandy. The position took Cuvier for the first time to France. Living in Caen, a lively cultural center, he was by no means isolated intellectually, and since his tutoring duties were not onerous he had plenty of time for natural history.

The outbreak of the French Revolution in 1789 hardly affected him at first; like many others of his generation throughout Europe, he was enthusiastic about the fall of the absolutist Old Regime. Later, however, as the Revolution lurched into its most radical phase, Cuvier witnessed scenes of atrocity that reinforced his profound horror of violence and social unrest, and his strongly rooted preference for firm government and social order and stability. As the political turmoil increased, his employers prudently withdrew from Caen to their country house near Fécamp. The revolutionary regime penetrated even to this rural fastness, though the d'Hericys were not harassed or persecuted. Cuvier even held an official position in the tiny community around their chateau: this was the first of many instances in which—following Lutheran tradition—he served the established political order, even when it was personally uncongenial. Although more isolated than in Caen, he still received scientific journals from Paris, and continued to correspond with some of his old friends and contemporaries in the German-speaking world.

Cuvier's first significant recorded geological comments came in two of the letters he wrote at this time to Christian Pfaff, a friend from his Stuttgart days (texts 1, 2).[2] These letters were not at all personal, and were probably handed round among his friends and others in Germany interested in the sciences, although they were not formally published until after Cuvier's death. They included detailed accounts of his own scientific activities; wide-ranging surveys of current work by others, as described in the scientific journals he was getting from Paris; and often lengthy reports, and Cuvier's comments, on current political affairs.

2. Christian Heinrich Pfaff (1773–1852) was later professor of medicine, physics, and chemistry at the University of Kiel.

In both letters the topics that would now be called geological were given the heading "mineralogy." That term still retained its traditional broad meaning, as one of the three primary divisions of "natural history": just as zoology and botany dealt with the description and classification of the animal and plant worlds, so mineralogy dealt with the same aspects of the inorganic terrestrial world. In the first letter, mineralogy was sandwiched between botany and zoology; in the second, between botany and chemistry. The texts give at least a glimpse of the way Cuvier was doing geological fieldwork, and even perhaps some mineralogical chemistry, in addition to wide reading of current scientific publications.

In the first letter (text 1), Cuvier described the geology of Normandy. While his remarks on the wider setting are likely to be derived from what he had read, his account of the Chalk plateau and coastal cliffs is clearly based on his own fieldwork there. The area is described in a style explicitly modeled on Abraham Werner (1750–1817), the already famous teacher at the great *Bergakademie* (mining school) at Freiberg in Saxony. Werner emphasized above all the importance of accurate observations in the field; in particular, rocks — or rather, rock masses or "formations" (*Gebirge:* literally, "mountains") — were to be described in terms of their three-dimensional or *structural* relations to each other and to the surface topography. Such rigorously descriptive work was termed "geognosy" (*Geognosie:* literally, "earth-knowledge").[3] However, Cuvier here went beyond normal geognostic practice, by tackling the *causal* problem of accounting for the deep valleys that cut through the thick horizontal beds of chalk. He argued that these particular valleys could hardly be due to any kind of tectonic disturbance, such as had often been suggested in other cases; yet he could not easily attribute them to erosion by streams and rivers, because many of them were "dry valleys" with no running water at all. The problem was left unsolved, but Cuvier's handling of it is quite impressive.

His analysis of another causal problem, that of the origin of the regular bands of flint nodules within the chalk, continued briefly in the following letter (text 2). Most of Cuvier's comments, however, were reserved for the high-level theorizing of Jean-André Deluc (or de Luc; 1727–1817), one of the most senior authors actively concerned with the study of the earth.[4]

3. In the nineteenth century, parts of geognosy evolved into the modern geological practice of stratigraphy, which is similarly descriptive in its primary orientation.

4. It would be anachronistic to call them geologists, for although the term "geology" had been coined — by Deluc himself — it was as yet rarely used, and its meaning was quite different from what it has since become (see below).

Deluc, a Genevan by birth, had settled in England, where he enjoyed a virtual sinecure at the court of King George III, which enabled him to spend most of his time in travel and writing. He saw himself as an Enlightenment philosopher, but, unlike most others, as a Christian too. His voluminous works on geology were designed not to create a new specialist science, but rather to deploy the widest possible range of scientific evidence to rebut the atheistic arguments of other philosophers. Deluc was no literalist in biblical interpretation, but he did believe that issues of the utmost importance to human life and faith were at stake in much current scientific work. In particular, he strove hard to demonstrate the historicity of Genesis, by showing how, when properly interpreted, it was compatible with the latest scientific observations. In Deluc's opinion, however, that did not entail constricting the history of the earth within the few thousand years of traditional chronology.

Cuvier was evidently already familiar with Deluc's earlier theorizing about the structure and history of the earth. But his letter is concerned with the current version of Deluc's theory, in which the scriptural significance of the science remained wholly implicit. Like much of his other work, Deluc's latest theory was unfolded in a leisurely series of "letters." This series was being published in monthly installments in the *Observations sur la physique, sur l'histoire naturelle, et sur les arts* (Observations on physics, on natural history, and on the arts), a leading scientific periodical edited in Paris.

Cuvier's synopsis of Deluc's work was fair and accurate; his evaluative comments were also remarkably mature for a twenty-two-year-old with, as yet, relatively little experience of scientific work. He clearly appreciated the value of trying to explain terrestrial phenomena in natural causal terms, as Deluc did, within a framework of earth *history;* yet he remained skeptical about the older naturalist's efforts to explain the whole range of phenomena within a single all-embracing theory. The conclusion of his review hinted at a reason for doubting whether this was the best way to proceed in the earth sciences: Deluc's theory was only one of several currently on offer, and this sheer plurality might be self-defeating.

Deluc's work should in fact be regarded as a typical example of a kind of scientific writing that was currently in vogue; it belonged to a literary tradition that reached back to Descartes. Borrowing from its first full expression, Thomas Burnet's *Telluris theoria sacra* (Sacred theory of the earth, 1681–89), this kind of writing was commonly termed "theory of the earth." Rather than referring to any particular theory, the phrase denoted a genre, indeed almost—in modern terms—a scientific discipline. It was in fact explicitly for this kind of work that Deluc had earlier proposed the name "geology." Any *particular* theory within this genre was usually referred to as a "system."[5]

"Theory of the earth" was concerned above all to explain the major features of the structure and history of the earth in terms of *natural causes.* Furthermore, those features were to be explained, as far as possible, in terms of just a few fundamental causes, rather than a profusion of different processes. The goal was a high-level theory that would share the simplicity of, say, Isaac Newton's explanation of complex planetary motions in terms of universal gravitation. "Theory of the earth" was not unconcerned with empirical observations, any more than Newton could have achieved his interpretation without the detailed observation of planetary orbits. But all empirical materials were deployed in the service of the overarching theory.

As Cuvier noted near the end of his review, he had focused on Deluc's grand theoretical "system," but that tireless traveler had backed it with a mass of detailed field observations. Those were by implication valuable in their own right, although not easily summarized. Cuvier referred to some of them as "geognostic facts"; that is, they were observations on the structure, sequence, and mineralogy of rock formations. Those who practiced geognosy contrasted that approach with what Deluc had termed "geology": the former was always rigorously observational and empirical, usually detailed and local, and primarily descriptive rather than causal in its aims; the latter might cite much observational evidence in its support, but its goal was high-level causal explanation.

In the closing years of the eighteenth century, these were two distinctive traditions in the scientific study of the earth. Cuvier's comments on Deluc's work show that he was already well aware of both. Their relation to each other, in effect the relation between observations and causal explanations in geology, was to be a central theme in all his work in that science.

---

5. It is worth noting here, at the start of this book, that the *Theory of the earth* (1795) by the Scottish philosopher James Hutton (1726–97), although now revered by anglophone geologists as one of their founding documents, was regarded by Cuvier and his contemporaries as just one among the many "systems" currently on offer, and not obviously superior to the rest (see text 19, sec. 20). In particular, Hutton's assumption that virtually unlimited time had been available for the operation of terrestrial processes was taken for granted by many others writing in the same genre. More distinctive—but poorly substantiated by empirical evidence—were Hutton's claims for the dynamic character of the earth's interior, and for its power to elevate whole continental masses out of the ocean floor. All in all, Hutton's status as an iconic "founder of geology" hardly stands up to modern historical scrutiny—but this is not the place to substantiate that assertion.

TEXT 1

⊢———⊣

Mineralogica

WERNER'S METHOD IS NOT unknown to me, although I haven't yet been able to buy his book.[6] On the subject of formations *[Bergen]* of chalk and flint, I'll give you a general idea of the country where I'm now living. This will not be useless to you, because it's quite different from anything in your vicinity, and will convince you how little foundation there is for all that is fantasized *[fabelt]* about the transformation of chalk into flint or of flint into chalk.

In geological terms, Normandy can be divided into three parts. The middle [part] around Caen and Bayeux is very low-lying, and the coast shades so gradually into the sea that there is a very wide difference between the boundary at high tide and at low; the sea therefore penetrates more deeply into the interior than in the other two parts. The western [part] from Cherbourg to Brittany is hilly and granitic, and is linked with Brittany [itself], which is of a similar nature. The eastern [part], to the north of the Seine, is by contrast nothing but a low plateau of chalk. This general arrangement corresponds exactly to that of England, which shows clearly that the two countries were separated from each other in the past. The similarity becomes still more striking if one thinks of the tin mines in Brittany and at the same time recalls that it is precisely in Cornwall, which corresponds to Brittany, that the most famous tin mines in the world are found.

The whole area around Caen consists, under the vegetable soil, of thick beds *[Schichten]* of a very compact sandy limestone, which is excellent for building and from which one can extract blocks as large as one wants. By contrast, in all the country that lies to the north of the Seine, perhaps no stone a rod long can be found, so the houses are built only of brick, flint, wood, or earth. This country is about 300–400 feet above sea level and is almost flat. But it is cut through by many quite narrow valleys, of which the smaller lead into the larger as if there were rivers in them; but at present there are streams only in the deepest valleys, and they are so very small that none to the north of the Seine has a name of its own. They are just named the stream of such-and-such a place, so that one of them can sometimes

6. [Probably Werner, *Kurze Klassifikation* (Short classification, 1787), the highly influential booklet in which the various rock "formations" *(Gebirge)* were classified primarily by their relative structural position, and subordinately by their relative age.]

have ten names along its course, from the different villages built beside it; most of them flow directly into the sea. But in the flat part of the country [i.e. on the plateau] there are neither streams nor springs, and the inhabitants have no other water than from the rainfall, which they collect in large cisterns dug for the purpose. The whole surface of the land is covered by a layer several feet thick of very fine reddish sand, mixed with clay; under the sand is the chalk, which always alternates with beds of flint, for as far from the coast as one can see.

At the sea's edge these alternating beds are cut perpendicularly, which yet again proves the former existence of a connection with England; but many other [facts] also show a close connection. If one looks at a place where such a valley emerges, for example at the harbor at Fécamp, one notices that the beds on one side correspond exactly to those on the other, in thickness, position, and nature [fig. 2]. If therefore they were once united, how have they been separated? This cannot have happened by any rupture, nor by a complete subsidence of what was in the middle, nor by a simple flexing. In the first case the sides of the valley would be steep. The second case is impossible, because the valley is narrower below than above, so that the upper beds could not have collapsed past the lower. The third [case] is also no good, for then the ends of the beds would be sharply inclined. No other means therefore remains, except flushing out the sludge with water. But the bed of sand was deposited only when the valleys were already formed, for it covers the slopes as much as the upper parts of the country, and lies very regularly on it in layers of various colors.

I now come in particular to the flints. They are in lumps of very irregular form, mostly of a brownish-black color, and from an inch to a foot in size; some are larger, and I've seen one that was three feet long. They lie in horizontal layers *[Schichten]* between the layers of chalk, and all the spaces between the flints are filled with chalk. In the vicinity of Fécamp I've

FIGURE 2　Cuvier's sketch of the geology of the chalk region of Normandy, showing a village built in a valley excavated in the horizontal beds of chalk and flint; note also the superficial sandy layer. This is probably a *view* of the coastal cliffs at Fécamp, as seen from offshore, not an imagined section through the land.

counted more than fifty of these alternations between chalk and flints. The flint layers are about a foot thick; the chalk layers from 3 to 4, 5 or 6 feet. I've not yet seen what some writers say, that the flints are like nests within the chalk layers.

Let me now say what is to be thought of the alleged transformation of chalk and flint, which Deluc has again maintained recently.[7] Does the air do it? Then all that is exposed to it would be transformed equally, and the perpendicular [cliff] face of the rocks would be uniformly either flint or chalk. Does some moisture circulating through the whole mass do it? How is it then that it works in such regularly horizontal planes? How is it that new layers or at least new lumps of flint aren't formed sometimes? Above all, I don't understand why people are so worried about this. If the chalk formation *[Kreideberge]* was deposited in the global sea that then covered the earth, why couldn't the chalk and flint material have been precipitated alternately? One often finds alternating beds of clay and sand—but enough of this.

The chalk can't be used for building, because it softens in the air; but it makes an excellent marl for improving the soil. Flints on the other hand are used for building houses; and this shows how false is the opinion that they weather in the air. It's only that most of those found in the fields still have the chalky crust that they had in their original position.

Extract translated from Cuvier to Pfaff, October 1791, printed in Cuvier, Briefen an Pfaff *(Letters to Pfaff, 1845), pp. 245–48. The style used in translating this and the following text is more informal than the others, to remind the reader that they come from letters not intended for publication.*

TEXT 2

⟨⟩

Mineralogy

A FEW WEEKS AGO I went down into a marl pit about 150 feet deep (I've already told you how chalk is used around here as marl), and had the opportunity to observe that the chalk and flint beds are arranged in the inland hills just as they are on the coast (this pit is 1½ hours [i.e. on foot] from the sea).

7. [As already mentioned, Deluc was in the middle of publishing a long series of "letters" (1790–93) in *Observations sur la physique*, on a variety of geological topics; a recent installment had dealt with chalk and its flints.]

With regard to the transformation of rocks I find in Brugnatelli's *Annals of chemistry* that the Academy [of Sciences] of St. Petersburg has confirmed the alteration of chalk into flint and that of gypsum into chalcedony. Has Wiedemann in his paper repeated Baumé's experiment? I'm well aware that Baumé is not a very reliable author, but still he claims so boldly to have altered flint into alumina that I would at least have tried his method.[8]

Although you have already mentioned Deluc and his major catastrophes *[Hauptcatastrophen]* several times, I think you only know of the earlier system he expounded in his *Physical and moral letters*.[9] He has made some major changes and as it were built a new system, the outlines of which are expounded in letters to Mr. de Lamétherie.[10] There are already 18 of these letters, but owing to a mass of digressions onto other subjects it is not yet complete. I hope it'll not be unacceptable to you, to find a sketch of it here.

In the introduction he sets out the major facts *[Hauptfacta]* on which his opinion is based; I'll omit these, since they are known to all mineralogists. He then establishes three major principles *[Hauptgrundsätze]*: (1) the stratification of the materials of our earth proves that they have all been deposited in a liquid; (2) their solidity, that they are the product of a crystalline precipitation; (3) their diversity, that successive changes occurred in the fluid, and these can only have been caused by new solutions or by the emission of new kinds of gas.

Now he divides the different events into six major periods *[Hauptperioden]*.[11] The first began when light penetrated the mass of our earth, combined to form heat *[Feuer]*, which made the water fluid. The core of the earth then had three layers: (a) the interior, to which water did not penetrate; (b) the middle, softened by water; and (c), all the surrounding liquid,

---

8. [The *Annali di chimici e storia naturale* (Annals of chemistry and natural history) were edited in Pavía by Luigi Gasparo Brugnatelli (1764–1818), the professor of chemistry there. Christian Rudolph Wilhelm Wiedemann (1779–1840) taught anatomy and surgery at the medical college in Braunschweig. Antoine Baumé (1728–1804) was professor of chemistry at the college of pharmacy in Paris. The individuals cited exemplify the cosmopolitan scientific network that Cuvier aspired to join.]

9. [Deluc, *Lettres physiques et morales* (1779). It was in a preliminary volume of this work (*Lettres sur les montagnes*, 1778) that Deluc had first proposed the term "geology," but as a synonym for "theory of the earth." "Physical" topics were those in any of the natural sciences; "moral" ones covered what would now be termed the social or human sciences.]

10. [Deluc, "Lettres à M. de la Métherie" (1790–93), published in *Observations sur la physique*. Jean-Claude de Lamétherie (or de la Métherie, or Delamétherie; 1743–1817) was the editor of the periodical in Paris.]

11. [This may have been an implicit reference to the six "days" of the creation narrative in Genesis. A sixfold periodization had also been adopted by Buffon (see below) in his "Époques de la nature" (1778).]

which contained the materials from which our atmosphere and the beds
[Schichten] of our continents have been formed.

Second period. Granite and similar kinds of rock precipitated. Limestone
is already found at this time.

Third period. Now come the primordial beds [Primordialschiefer], which
contain no organic substances. Thus a solid crust formed around the soft
layer (b); water filters through into the soft underlying layer and takes part
of it with it. (Here the system shows itself to be very weak: why is this crust
solid, since it is also formed in water? how can water take the soft part with
it, and where, since everything is full up?) I continue with Deluc's story
[Erzählung]: as a result of part of the soft layer being carried away, huge
cavities form; the parts of the overlying crust that constitute the roofs of
these cavities collapse; the crust breaks across the still solid edges between
the caverns; so the parts nearest the fractures stand on edge and form the
primordial mountains, which are left dry by the water that has now sunk
into the caverns. Coal proves that vegetation must have existed immediately after this.

Fourth period. Gases [expansible Fluida] develop in the caverns, break
out through cracks in the crust, and throw fragments of the deeper layers
on to the uppermost; these are consolidated by new precipitations. Hence
the primordial breccias.

Fifth period. The ejected gases and the influence of the sun change the
chemical nature of the liquid that still covers almost the whole earth. This
changes the nature of the precipitation, and the primordial limestones are
formed. Marine animals begin to show themselves. Meanwhile water fil-
tered into the interior caverns as before; this likewise led to a collapse,
which broke the limestone crust just as the granitic had been broken earlier. Hence all granitic chains have on either side a slaty and, somewhat
further, a limestone chain, which are separated by two parallel valleys.[12]
After this second collapse there were new infiltrations and emissions of new
kinds of gas, so that new precipitations occurred once more, but not so
generally and of a quite different kind. One notices particularly (1) a second limestone formation, in which marine animals are very abundant; (2) a
less widespread sandstone formation, which also contains marine animals,
but mostly of other genera than in the preceding ones; (3) these two formations suffered further, but merely local, catastrophes; (4) new eruptions of
gases threw on to this layer pieces of the preceding ones. During all these

12. [Mountain ranges were generally believed to be structured symmetrically around a core of
granitic rocks, flanked by zones of overlying and therefore successively younger rocks.]

revolutions it often happened that dry land, on which vegetation had existed for many years, sank again and was covered with new beds of stone.
Hence the Coal beds. Finally came the Chalk beds;[13] they contain no more
ammonites, but on the other hand they have other marine animals that are
not found today, for example the sea urchin of which the spines are called
jew-stones [Judensteine]. It should also be noted that the Chalk has only
been deposited in certain places.

Thus far has Deluc got at present. He has promised to deal next with the
sixth period, in which our continents were put into the state in which they
still are today.[14] His opinion is subject to great difficulties, but these letters
are very interesting on account of many very firmly handled points of
physics and numerous geognostic facts [geognostische Facta]. Admittedly it
is just this part that is not suitable to be reviewed here.

De Lamétherie himself wants to contribute to this too, and also proposes
a new system; but I can tolerate de Lamétherie's method of reasoning as
little as his style. A third system, by Father Pini of Turin, was published some
months ago, but I know of it only from reviews.[15]

Extract translated from Cuvier to Pfaff, 11 March 1792, printed in Cuvier, Briefen an Pfaff
(1845), pp. 257–60.

13. [Deluc, like Werner and many others, regarded the distinctive Chalk as the *uppermost* of all
the regularly bedded Secondary formations mentioned in this passage, overlain only by loose superficial deposits such as the clay Cuvier had described in Normandy (text 1). This is why Cuvier's
much later research around Paris, in which the Chalk featured as the *lowest* of many distinct formations, was in its time so strikingly novel (see chapter 12).]

14. [It is unfortunate that Cuvier did not write this letter a few months later, for we do not have
his immediate reaction to Deluc's "sixth period," the part of the "system" that was most important
for Cuvier's own later geological views. In the twelve letters published after Cuvier wrote this summary, Deluc duly set out his arguments for claiming that the continents—in their present form as
landmasses—were not of great age, because they had emerged from the ocean floor only a few thousand years ago. (Deluc also dealt with matters as diverse as the rings of Saturn, chemical affinities,
and gravity itself!)]

15. [De Lamétherie published his own "system" later, as *Théorie de la terre* (1795; 2nd ed. 1797).
Ermenegildo Pini (1739–1825) was professor of natural history at the University of Milan and a
member of the Barnabite teaching order; his *Rivoluzione del globe terrestre* (Revolutions of the terrestrial globe, 1790–92) had in fact been criticized by Deluc in an earlier letter published by de
Lamétherie.]

2

LIVING AND FOSSIL ELEPHANTS

In 1795, three years after Cuvier told his friends in Germany about Deluc's latest theory, the political situation in Paris became more stable, or at least more favorable for scientific work. During the Terror, the most radical and violent phase of the Revolution, many of the old institutions of science had been abolished, or at least disrupted. Many of the most influential savants had fled from the capital.[1] Some, most notably the great chemist (and tax collector) Lavoisier, had even lost their lives at the guillotine. Now yet another coup d'état had given France a politically more moderate government, the so-called Directory, which quickly showed itself more favorable to the sciences than any since the start of the Revolution.

Cuvier therefore made a bold and risky decision to move to Paris in search of a scientific career. In this he was encouraged by meeting a scientific refugee from the capital, who wrote to colleagues there on his behalf. Cuvier had already sent some articles (on invertebrate zoology) to be published in Paris, but he was still scarcely known, and had no certainty of gaining any position. In the event, however, he could hardly

1. The contemporary term "savants" (which was used in English as well as in French) will be used throughout this volume, in place of the misleadingly anachronistic term "scientists." Savants could be learned, expert, or "savant" in any of a wide range of subjects, not just those covered by the modern anglophone meaning of "science"; and they might or might not be "professionals" in the sense of earning their living from such studies.

FIGURE 3   A portrait of Georges Cuvier at the age of twenty-six—possibly a self-portrait—drawn in 1795, around the time he moved to Paris; it may have been made to further his career prospects.

have arrived at a more propitious time. As a result of the Terror, the old networks of patronage that had been essential for making a career in science had been thrown into disarray, and had yet to be reconstituted: a young man of talent had more opportunities than ever before (fig. 3).

Given Cuvier's interests, it is not surprising that he focused his attention on the Muséum National d'Histoire Naturelle (National Museum of Natural History). Almost alone among the major scientific institutions in Paris, this had escaped abolition, because at the height of the

Revolution it had reformed itself in a politically correct manner. Although new in name, it was in fact the direct successor of the old royal botanical garden (Jardin du Roi) and the associated royal museum (Cabinet du Roi). Here at the new Muséum,[2] Cuvier managed—not without opposition—to obtain a junior position as understudy *(suppléant)* to Mertrud, the elderly and undistinguished professor of animal anatomy. The Muséum was to be Cuvier's professional home, and, before long, his domestic home too, for the rest of his life.

Even a modest position at the Muséum placed Cuvier at the world center for the natural history sciences, and its incomparable collections became at once his most important resource. Before the end of the year, his lecture course on comparative anatomy at the Muséum (standing in for his nominal superior) showed Parisian savants that he was a newcomer to be reckoned with. He put his science firmly on the map, by explaining his conception of organisms—though it was not original to him—as functionally integrated "animal machines."

A few weeks earlier, in one of its major acts of cultural politics, the Directory had approved the foundation of a new Institut National. This was intended to repair the revolutionary damage to French science and scholarship, by bringing together in one prestigious body all the branches of knowledge formerly cultivated in the various learned "academies" that had been suppressed. Among these was the old Académie Royale des Sciences (Royal Academy of Sciences), which was in effect revived as the Institut's "class for mathematical and physical sciences." Significantly, it was termed the *First* Class of the Institut (in modern terms the three classes covered, roughly and respectively, mathematics and the natural sciences, the social sciences, and the humanities).

Only a week after Cuvier's inaugural lecture, and doubtless partly as a result of that event, he was elected the youngest member of the First Class. Just as the Muséum became the site of his actual research, so the Institut became the main arena for the exposition of his scientific results, as several of the texts in this volume show. Cuvier's rise to prominence in Parisian science in the years that followed continued to be meteoric, but it was not effortless. Like any scientific career in this period, it required the painstaking construction of networks of patrons and allies, and discreet campaigns against rivals on all sides.

Once installed in the Muséum, however precariously at first, Cuvier picked up the research on comparative anatomy that he had started in

---

2. The accent and initial capital will serve hereafter to indicate reference to this specific museum—at the time, the greatest natural history museum in the world.

Normandy. He began to produce important papers on the anatomy of the then poorly understood marine invertebrates, particularly the mollusks. But the resources of the Muséum quickly turned his attention to the vertebrates too, and above all to the mammals. More specifically, he soon saw that some recent acquisitions to the Muséum's collections might make it possible to settle a long-standing problem with far-reaching implications.

It had long been known that large fossil bones and teeth were found widely scattered in northern latitudes, in both the Old World and the New, in "superficial" deposits close to the surface of the ground. They were far from the tropical habitats of all the known large mammals such as elephants and rhinoceros. The identification of these fossil bones, and the explanation of their anomalous geographical position, had long been matters of lively international debate among naturalists.[3] Louis Jean Marie Daubenton (1716–99), now the professor of mineralogy at the Muséum and one of Cuvier's senior colleagues, had been a major contributor to this debate before the Revolution; and George Louis Leclerc, count de Buffon (1707–88), for almost half a century the director *(intendant)* of the Muséum's forerunner, had made the fossil bones a key component in his overarching "theory of the earth." So Cuvier was entering a well-trodden field.

He had one major empirical advantage over his predecessors. Among the incidental spoils of the revolutionary wars were the outstanding collections of the former ruler of the conquered Netherlands. What had recently reached Paris included not only paintings and other items of great artistic importance, but also a major natural history collection. It included specimens that, added to those already at the Muséum, proved to Cuvier's satisfaction that the living African elephant was not the same species as the Indian, as had been commonly supposed; and that the *fossil* elephant or "mammoth" was anatomically distinct from either. Cuvier was not the first naturalist to suspect this; but he alone had both the means and the skill to demonstrate it persuasively.

Just a year after his arrival in Paris, he presented his first paper to the Institut, setting out this argument. A summary of the paper (text 3) was published soon afterward in the *Magasin encyclopédique* (Encyclopedic magazine), a newly founded journal for all the sciences, which took its inspiration from the great French *Encyclopédie,* the supreme emblem of the eighteenth-century Enlightenment. The full version was published three

---

3. "Naturalists" was the contemporary term for those who studied the sciences of "natural history" such as zoology, botany, and mineralogy; neither term had its modern pejorative overtones of amateurism. "Naturalists" were, in effect, a subset of the larger category of "savants."

years later in the Institut's new *Mémoires*, with several plates of engraved illustrations based on his own drawings of the crucial evidence (see fig. 4).[4]

Cuvier's first major paper displayed remarkable self-assurance—some might term it arrogance—for a twenty-six-year-old with little scientific achievement to his name. Emphasizing the importance of a critical evaluation of factual claims, he confidently rejected the opinions of his distinguished predecessors, on the grounds that their observations had been insufficiently precise. He presented his conclusions about the three distinct species of elephants as a triumph for his own scrupulously exact methods of osteological comparison. Almost in passing, he dismissed any suggestion that the differences might be due to the transformation (in modern terms, evolution) of one species into others—a notion that in general terms was being actively canvassed in Paris at this time—and maintained that to abandon the concept of the stability of natural species would be to subvert the whole taxonomic enterprise. But he was careful to argue that his anatomical approach could only enrich and deepen the traditional zoological emphasis on the externally visible characters of animals. This related his work tactfully to that of an even more youthful colleague, the professor of zoology Étienne Geoffroy Saint-Hilaire (1772–1844), who had helped him gain his position at the Muséum.

Cuvier also presented his work as a demonstration of the way comparative anatomy could be an ancillary but essential tool for establishing the "theory of the earth," or "geology," on less speculative foundations. He argued that his research had undermined the impressive edifice of the celebrated theory of the earth that Buffon had expounded in his "Époques de la nature" (Epochs of nature, 1778). This had been centered on the idea—not original to Buffon—that the earth had had its origin as an incandescent body in space, and that it had cooled gradually to its present surface temperature. Buffon had assumed that the bones found in northern lands were those of elephants and other tropical species, and had therefore used them as evidence of a formerly warmer climate at high latitudes. But if, as Cuvier now argued, the mammoth was not the same species as either of the living elephants, it could well have been adapted to a quite different environment, namely to the cold climates in which its bones were now found; Buffon's argument for a cooling earth, or at least his use of the bones as evidence for it, would then collapse.

Cuvier's inference left new problems, however, above all that of accounting for the difference—as he claimed it to be—between *all* the

4. Cuvier had shown outstanding talent as a biological artist even in his youth; he continued throughout his life to make most of his own drawings, though they then had to pass through the hands of professional engravers before publication.

known fossil species and those now alive. In fact, when he first presented his paper, he made his claim even more sweeping than it appeared in print, because he extended this absolute contrast between fossil and living species to *marine* animals as well as such terrestrial species as the elephants. But after his lecture the "learned conchologists" he had cited must have rejected that claim, insisting that some marine mollusks did have exact "analogues" among fossil shells.[5] Even with an implicit restriction to terrestrial animals, however, his published claim was striking enough.

Cuvier claimed—though without detailed argument—that the evidence pointed to an earlier and *prehuman* "world" that had been "destroyed by some kind of catastrophe." This was a theme that, though not original to him, was to pervade his geological theorizing for the rest of his life. Although he did not explain why the event must have been sudden, he did imply that it was not unique, and that it might be repeated in the future. But he deftly drew back from further speculation of this kind, leaving such matters to a bolder—or perhaps more foolhardy—"genius." This was a neat way of deferring, though with more than a touch of irony, to his senior colleague Barthélemy Faujas de Saint-Fond (1741–1819), who had boldly adopted Deluc's neologism "geology" as the title of his professorship when the Muséum was reconstituted.

TEXT 3

———

Memoir on the Species of Elephants, Both Living and Fossil

Read at the public session of the National Institute on
15 Germinal, Year IV [4 April 1796][6] by G. Cuvier[7]

CONSIDERABLE DIFFERENCES have long been noted between the elephants of Asia and those of Africa, with regard to their size, their habits,

---

5. The relevant passage in the manuscript (MS 628, Bibliothèque Centrale, Muséum National d'Histoire Naturelle, Paris) was omitted from the first published text of the paper (translated below as text 3) and its subsequently enlarged versions: see Burkhardt, *Spirit of system* (1977), p. 129 and n. 56.

6. [The date, like several others in this volume, is given in the form of the Republican calendar. This was introduced at the height of the Revolution as part of the effort to eliminate all traces of the culturally Christian past. It had its nominal origin at the declaration of the French Republic in September 1792 (though it was not introduced until Year II, or 1793–94); it divided the year (beginning in September) into twelve new months based on the seasonal weather. The Republican calendar was dropped, and the ordinary (Gregorian) calendar resumed, at the start of 1806. (In this volume Republican years will be denoted by Roman numerals, as they often were at the time.)]

7. This article is an abstract of a detailed paper that will be printed in the Institute's collection,

and the places where they live; and Asiatic peoples have known since time immemorial how to tame the elephants they use for hunting, whereas African elephants have never been subdued, and are hunted only to eat their flesh, to collect their ivory, or to eliminate the danger of their presence. Nonetheless the authors who have dealt with the natural history of elephants have always regarded them as forming one and the same species.

The first suspicions that there are more than one species came from a comparison of several molar teeth that were known to belong to elephants, and which showed considerable differences; some having their crown sculpted in a lozenge form, the others in the form of festooned ribbons.

The arrival in Paris of the natural history collection acquired for the Republic by the Treaty of The Hague has enabled us to turn these suspicions into certainty.[8] It contains two elephant skulls: one, which has the teeth with festooned ribbons, comes from Ceylon; the other, which has only diamond forms, is from the Cape of Good Hope. A glance at these skulls is sufficient to observe, in their profile and all their proportions, differences that do not allow them to be regarded as the same species (fig. 4). It is clear that the elephant from Ceylon differs more from that of Africa than the horse from the ass or the goat from the sheep.[9] Thus we should no longer be astonished if they do not have the same nature or the same habits.

It is to anatomy alone that zoology owes this interesting discovery, which a consideration of the exterior of these animals would only have been able

---

accompanied by the necessary descriptions and illustrations [Cuvier, "Espèces d'éléphans" (Species of elephants, 1799)].

8. [The treaty established the terms of peace between the victorious French and the Dutch they had defeated. As part of the officially sanctioned cultural looting of the Netherlands, the fine natural history collection of the Stathouder, the Dutch ruler who had fled to England, was removed to the Muséum in Paris.]

9. [The full text of the paper has at this point one of Cuvier's most trenchant statements of his rooted opposition to evolutionary interpretations of organic diversity: "I believe that, after reading this comparative description, which I have made with all possible care and precision, and for which the original specimens exist in the comparative anatomy collection at the Muséum, no naturalist can doubt that there are two quite distinct species of [living] elephants. Whatever may be the influence of climate to make animals vary, it surely does not extend this far. To say that it can change all the proportions of the bony framework [charpente osseuse], and the intimate texture of the teeth, would be to claim that all quadrupeds could have been derived from a single species; that the differences they show are only successive degenerations: in a word, it would be to reduce the whole of natural history to nothing, for its object would consist only of variable forms and fleeting types [types fugaces]" (1799, p. 12). The word "dégénérations" was widely used to denote changes within a species, forming some new variety; but also, by at least some authors, for changes transforming one species into another.]

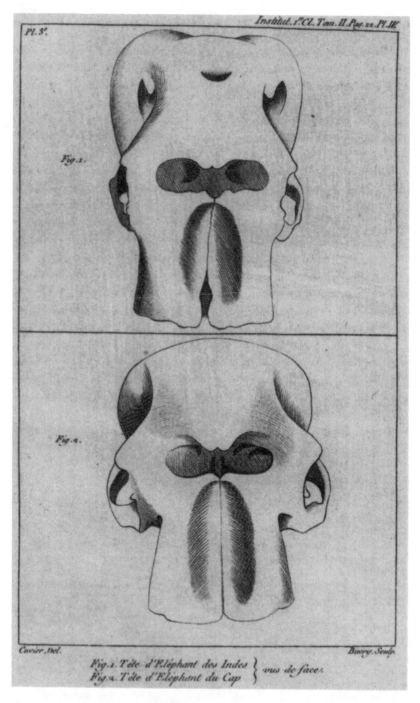

FIGURE 4   The skulls of elephants *(top)* from Ceylon (now Sri Lanka), south of the Indian mainland, and *(bottom)* from the Cape of Good Hope (now in South Africa), engraved from Cuvier's drawings and published in 1799 with the full text of his paper.

to render imperfectly.[10] But there is [also] a science that does not appear at first sight to have such close affinities with anatomy; one that is concerned with the structure of the earth, that collects the monuments of the physical history of the globe, and tries with a bold hand to sketch a picture of the revolutions it has undergone:[11] in a word, it is only with the help of anatomy that *geology*[12] can establish in a sure manner several of the facts that serve as its foundations.

Everyone knows that bones of enormous animals are found underground in Siberia, Germany, France, Canada,[13] and even Peru, and that they cannot have belonged to any of the species that live today in those climates. The bones that are found, for example, throughout the north of Europe, Asia, and America resemble those of elephants so closely in form, and in the texture of the ivory of which their tusks are made, that all savants hitherto have taken them to be the same. Other bones have appeared to be those of rhinoceros, and they are indeed very similar: yet today there are elephants and rhinoceros only in the tropical zone of the Old World. How is it that their carcasses are found in such great numbers in the north of both continents?

On this point, one is left with [mere] conjectures. Some [writers] have invoked great inundations that have transported them there; others suppose that southern peoples led them there in some great military expeditions.[14] The inhabitants of Siberia believe quite simply that these bones come from

10. [Cuvier and other naturalists of his generation were critical of the zoology practiced by their predecessors (e.g. Buffon) for having focused attention on the externally visible characters of animals rather than the internal anatomy revealed by dissection. Cuvier himself was highly skilled in practical dissection; in this respect his studies of molluscan anatomy are even more striking than his work on vertebrates, since they involved much finer manual dexterity.]

11. [In Cuvier's writing and that of his contemporaries, the word "revolution" simply meant major changes in the course of time: it was used for example in the writing of human history to denote the slow rise and fall of civilizations; and in astronomy to denote the regular orbiting of the planets round the sun. It had no *necessary* connotations of suddenness, still less of violence. In effect, what Cuvier termed "catastrophes" (see below) were a special subset of "revolutions."]

12. [The emphasis is not indicated typographically in the original, but is implied by the construction of the sentence. It is important to remember that at this time the term was still a neologism that had been adopted by very few writers other than its author Deluc and Cuvier's colleague Faujas.]

13. ["Canada" included much of what eventually became the United States: in particular, the uncolonized country around the Ohio River, which yielded some of the most problematic fossil bones.]

14. [A huge mass of water sweeping suddenly across the continents (like the tsunamis associated with some submarine earthquakes, but far larger) was a widely favored explanation for the bones found in Siberia. The classical accounts of Hannibal's campaign from North Africa, complete with some military elephants, had been the basis for an earlier explanation of the fossil bones found in Europe, but its plausibility had collapsed as more and more bones were found.]

a subterranean animal like our moles, which never lets itself be taken alive; they name it "mammoth," and mammoth tusks, which are similar to ivory, are for them a quite important item of commerce.

None of this could satisfy an enlightened mind [un esprit éclairé]. Buffon's hypothesis[15] was more plausible, if we assume that it was not contentious for reasons of another kind. According to him, the earth had emerged burning from the mass of the sun, and had started to cool from the poles; it was there that living nature had begun. The species that formed first, which had more need of warmth, had been chased successively toward the equator by the increasing cold; and since they had traversed all the latitudes, it was not surprising that their remains were found everywhere.

A scrupulous examination of these bones, made by anatomy, will relieve us of having recourse to any of these explanations, by teaching us that they are not similar enough to those of the elephant to be regarded as absolutely from the same species. The teeth and jaws of the mammoth do not exactly resemble those of the elephant [fig. 5]; while as for the same parts of the Ohio animal, a glance is sufficient to see that they differ still further.[16]

These [fossil] animals thus differ from the elephant as much as, or more than, the dog differs from the jackal and the hyena. Since the dog tolerates the cold of the north, while the other two only live in the south, it could be the same with these animals, of which only the fossil remains are known.

However, while relieving us of the necessity of admitting a gradual cooling of the earth, and while dispelling the gloomy ideas that presented the imagination with northern ice and frost encroaching on countries that today are so pleasant, into what new difficulties do these discoveries not now throw us?

What has become of these two enormous animals of which one no longer finds any [living] traces, and so many others of which the remains are found everywhere on earth and of which perhaps none still exist? The fossil rhinoceros of Siberia are very different from all known rhinoceros. It is the same with the alleged fossil bears of Ansbach;[17] the fossil crocodile of

15. [Buffon, "Époques de la nature" (1778). As a leading philosopher of the Enlightenment, Buffon was an "enlightened mind" par excellence.]

16. ["Ohio animal" referred to bones first found in 1739 on the banks of the Ohio River (in what is now Kentucky): their identity was much disputed during the rest of the eighteenth century, and was not resolved until Cuvier later defined and named the animal *Mastodon*.]

17. [The bones found in caves in a part of Bavaria that at this time was in the territory of Ansbach, most famously in caves around Muggendorf, between Erlangen and Bayreuth.]

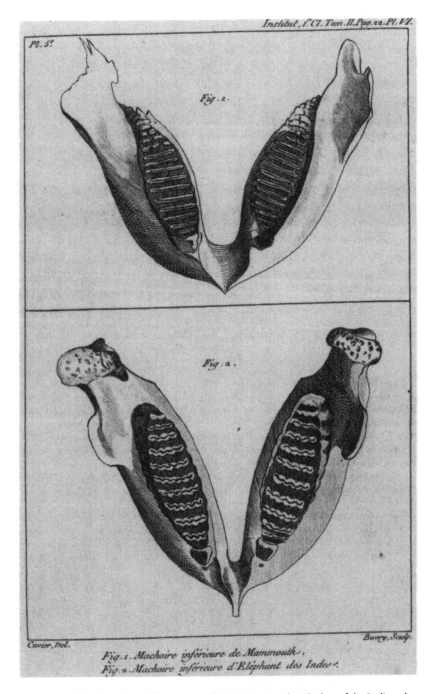

FIGURE 5 The lower jaw of the mammoth *(top)* compared with that of the Indian elephant *(bottom)*, engraved from Cuvier's drawings and published in 1799 with the full text of his paper.

Maastricht; the species of deer from the same locality;[18] the twelve-foot-long animal, with no incisor teeth and with clawed digits, of which the skeleton has just been found in Paraguay [see fig. 6]: none has any living analogue.[19] Why, lastly, does one find no petrified human bone?

All these facts, consistent among themselves, and not opposed by any report, seem to me to prove the existence of a world previous to ours, destroyed by some kind of catastrophe.[20] But what was this primitive earth? What was this nature that was not subject to man's dominion? And what revolution was able to wipe it out, to the point of leaving no trace of it except some half-decomposed bones?

It is not for us [i.e. Cuvier himself] to involve ourselves in the vast field of conjectures that these questions open up. Only more daring philosophers undertake that. Modest anatomy, restricted to detailed study and to the scrupulous comparison of the objects submitted to its eyes and its scalpel, will be content with the honor of having opened up this new highway to the genius who will dare to follow it.

Translated from Cuvier, "Espèces des éléphans" (Species of elephants, 1796).

18. [The "crocodile" was a spectacularly large fossil found in underground quarries near the southern Dutch town. The finest known specimen had recently been brought to Paris, like the elephant skulls, as a trophy of war. It was described and illustrated in a lavishly produced monograph by Faujas, *Montagne de Saint-Pierre de Maestricht* (Saint Peter's Mount at Maastricht, 1799), which he must have been preparing at this time. It was later interpreted as a huge marine lizard, and Cuvier named it *Mosasaurus* (lizard of the Maas or Meuse) (see chapter 13). "Deer" referred to supposed fossil antlers from the same Chalk formation at Maastricht, which Cuvier—once he had seen the specimens—identified as parts of the carapace of a marine turtle.]

19. ["Analogue" was the term used in the contemporary debate about the reality or otherwise of extinction, to denote a living species that was *identical* to one found fossil. For the Paraguay animal, see text 4.]

20. [The full text of the paper has a significant addition at this point: "beings whose place has been filled by those that exist today, which will perhaps one day find themselves likewise destroyed and replaced by others" (1799, p. 21). For Cuvier the present "world" had no finality, and the "catastrophe" that had made the mammoth extinct was certainly not a unique event, and perhaps not even the last of its kind.]

3

THE MEGATHERIUM FROM

SOUTH AMERICA

———

Most of the fossils that Cuvier mentioned in his paper on living and fossil elephants had already been described and discussed by others; but one of them, as he noted, was a recent discovery. Cuvier made this the subject of a separate paper, which he read at the Institut National not long after his first. It greatly increased his personal stake in the field of *fossil* anatomy.

Fossil bones are usually found scattered and disarticulated. However, one almost complete assemblage of bones, clearly derived from a single individual of some large animal, had been found in 1789 near Buenos Aires in what was then Spanish South America. Shipped back to Madrid, these bones were assembled at the Gabinete Real (Royal Museum) by Juan-Bautista Bru (1740–99), a conservator there. The most important separate bones and Bru's mounted reconstruction of the whole skeleton were drawn and engraved for him in preparation for a paper he planned to write about it. In 1796 a French official who was visiting Madrid saw the skeleton and obtained a set of Bru's unpublished plates. These were sent to the Institut in Paris, and Cuvier was asked to report on them. In his paper, Cuvier claimed that the unknown animal was an edentate, and named it *Megatherium,* or "huge beast" (text 4).[1]

Cuvier went no farther in a geological direction than to conclude that the megatherium, like other fossil species, must surely be extinct. The paper is included in this volume because it shows the methods of careful anatomical comparison that underlay all Cuvier's geological inferences. Specifically, it illustrates his theoretical concept of the "subordination of characters," according to which the different functions of the animal body formed a kind of natural hierarchy, such that some anatomical features were more reliable than others, for assessing the natural affinities between any one animal and other related forms. The application of this principle, in this case, underlay Cuvier's confident conclusion about the place of the fossil mammal from South America, in relation to living mammals.

However, Cuvier needed practical skills and empirical evidence as well as biological theory. Only a handful of naturalists anywhere had the skill and experience to understand the anatomy of the unfamiliar and exotic edentates, sufficiently for the case in hand; and only at the Muséum in Paris, probably uniquely at the time, could *any* naturalist have found the range of rare specimens necessary to establish by osteological comparison that the huge megatherium was related to the lowly sloths and anteaters. It was a striking conclusion.

The megatherium itself remained in Madrid, but Cuvier's paper—published in the *Magasin encyclopédique* shortly before the one on elephants—made the fossil widely known, particularly since it was accompanied by a crude copy of Bru's engraving of the skeleton (see fig. 6).[2] As an elephant-sized animal quite unlike any living species, it was a sensational addition to the growing collection of large vertebrates that—Cuvier claimed—could not plausibly be supposed to be still alive anywhere on earth.

---

1. The episode has been the subject of much chauvinistic argument. Rather than feeling that Cuvier had upstaged him, Bru may have valued the French naturalist's authoritative report on the zoological affinities of the animal. Conversely, Cuvier knew almost nothing about its geological context, as the text of his paper shows. Bru's plates were published in Spain later the same year, with his detailed description of the find, and a translation of Cuvier's paper; conversely, when Cuvier came to publish a full version of his own paper, he included a translation of Bru's work.

2. This engraving was copied in turn for the English *Monthly magazine,* and published later the same year with a summary of the paper; among the anglophone naturalists who thus became aware of the megatherium was Thomas Jefferson (1743–1824), not only a prominent politician in the young United States but also a keen naturalist who had already studied similar bones from Virginia. The Spanish translation of Cuvier's paper has been noted already.

TEXT 4

*Note on the skeleton of a very large species of quadruped,
hitherto unknown, found in Paraguay and deposited in the Cabinet
of Natural History in Madrid. Edited by G. Cuvier.*

THIS SKELETON is fossil: it was [found] a hundred feet below the surface
of a sandy formation *[terrain]* near the river La Plata.³ It lacks only the tail
and some paired bones that have been imitated in wood. It is mounted in
Madrid, where Citizen Roume, correspondent of the National Institute, has
examined it carefully. The complete figure and all the details have been
engraved on five plates in folio format, which are probably intended to illustrate some dissertation of which this skeleton will be the object. The National Institute having received proofs of these plates from Citizen Grégoire, they have served as the basis for the present note, together with a
short description sent by Citizen Roume.⁴

This skeleton, shown in [fig. 6], is twelve feet long and six in height. The
backbone is composed of seven cervical vertebrae, sixteen dorsal, and four
lumbar; there are therefore sixteen ribs. The sacrum is short, the iliac bones
very broad; and their plane being almost perpendicular to the spine, they
form a wide-open pelvis. There is no pubis or ischium at all, or at least they
are lacking in this skeleton, and one can see no mark where they would
have been during the life of the animal.

The thigh bones are extremely thick, and those of the legs still more so in
proportion. The entire sole of the foot was on the ground when walking. The
shoulder blade is much broader than long; there are perfect clavicles, and
the two bones of the forearm are separate and movable one on the other.
The forelimbs are longer than the rear. Judging by the form of the last phalanges, there must have been very large pointed claws, covered at their base
in a bony sheath; it appears that there were only three of these claws on the

3. [It was found at Luján, west of Buenos Aires (in modern Argentina). The "Paraguay" of the title
was a mistake, which indicates how little Cuvier knew about the circumstances of the find. The
name stuck, and the fossil continued for several years to be called "the Paraguay animal."]

4. ["Citizen" was the egalitarian title that had been imposed at the height of the Revolution, to
replace all the subtly nuanced forms of address used under the Old Regime. Cuvier's claim that he
knew nothing about the provenance or intended use of the plates was perhaps genuine; on the other
hand it may have been a covert way of establishing his own priority in the interpretation of the bones.
Henri Baptiste Grégoire (1750–1831) was a priest who had been prominent in Republican politics,
and later in setting up the Institut National and other scientific bodies. Philippe-Rose Roume had
been on a governmental mission to the French colony of Saint-Domingue (now Haiti).]

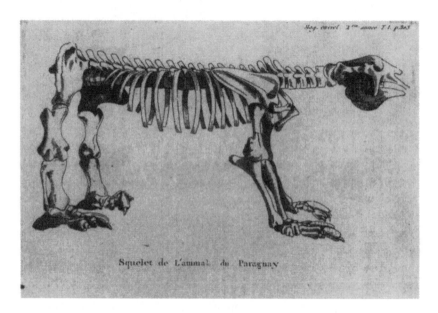

Squelet de L'animal du Paraguay

FIGURE 6 Bru's reconstruction of the skeleton from South America, as redrawn to illustrate Cuvier's paper on what he named the megatherium.

forefeet and a single one on the rear feet, and that the other digits lacked them and perhaps that they were entirely hidden under the skin.

The head is the most singular part of this skeleton. Its occiput is elongated and flattened, but it is fairly convex above the eyes. The two jaws form a considerable toothless projection, and have at the back of the mouth only four teeth on each side, both above and below, all of them molars, with a flat crown and channeled transversely. Above all one should notice the breadth of the sides of the lower jaw, and the large apophysis placed at the base of the zygomatic arch.

This animal differs, in the ensemble of its characters, from all known animals; and each of its bones, considered separately, also differs from the equivalent bones of all known animals. This is what results from a detailed comparison of this skeleton with those of other animals, and will easily be seen by all who are familiar with this kind of research; for none of the animals that approach this one in size have either pointed claws, or a form of head, of shoulder blades, of clavicles, of pelvis, or of limbs like those of this animal.

As for the place of this quadruped in the [natural] system, it is marked perfectly by a simple inspection of the ordinary *indicative characters;*[5] that

5. [That is, the features that most clearly reveal the natural affinities of an organism, locating it in a "natural" taxonomic classification, rather than in an "artificial" one designed purely for ease of identification.]

is to say the claws and the teeth. They show that it ought to be placed in the family of unguiculates lacking incisor teeth; and in fact it has striking affinities with these animals in all parts of its body. This family comprises the sloths *(Bradypus)*, the armadillos *(Dasypus)*, the pangolins *(Manis)*, the anteaters *(Myrmecophaga)*,[6] and the aardvark *[orycterope]* or Cape anteater.

The sloths and anteaters have claws exactly like those of our animal, borne in the same way on an axis, and encased at their base by a bony sheath; they have, like it, several digits that are obliterated and lack claws; such that it is among their species that one finds the least common arrangements, such as two digits in front and three behind, or two and four, or three and three, etc. Our animal also has a number of claws that is singular and indeed hitherto unique, namely three in front and a single one behind.

The greater length of the forelimbs is a character singularly specific to the sloth genus, but is much stronger among them than it is here, and it is the principal cause of their slow gait. In this respect, our animal is thus distanced a little from the sloth genus and approaches those that have greater equality between their extremities. The extraordinary thickness of the rear limbs is also found, to some extent, in the pangolin, which has the thigh and leg bones thicker in proportion to their length than any other animal, except this one.

The family of animals of which we speak presses on the heel when walking, as does this Paraguay animal. Most of its species have similar clavicles. If in fact the pelvis has no pubis or ischium at all, it is likewise in this family alone that we would find a faint trace of this anomaly. It is true that the two-toed sloth has these two bones, but they are not fused in front, and always remain separate. This same two-toed sloth has an arm bone wholly like this one, above all in the breadth of the lower part. Finally, it also resembles it in the thickness of the bone from the elbow toward the wrist, a fairly rare character among the quadrupeds.

As for the head, although it is very different from all known forms, it is nonetheless again in the family of edentates that one finds forms from which it is less distant than all the others; but, in order to grasp the relations better, it is well to give here a brief sketch of the forms of the head that this family show us.

The anteaters and pangolins have no teeth; their lower jaw, serving only to house the tongue, is slender and without any strength in its bones or in the muscles that close it; there is no coronoid apophysis at all, and the

6. [These four Latin names were printed as footnotes, and identified as being those of Carl von Linné (1707–88) the Swedish naturalist and leading taxonomist (better known as Linnaeus, from his publications in Latin).]

zygomatic arch is imperfect; the head itself is conical or even elongated
into a cylinder. This form is also that of the aardvark or Cape anteater, but
the latter is provided with molar teeth, and it feeds on roots; the jaw is
broad behind, and provided with a coronoid apophysis for the insertion of
the temporal muscle.

The armadillos have nearly the same kind of life as the aardvark and the
same form of jaws, with very similar teeth; their head is only a little shorter
and more pointed. In both genera the zygomatic arch is complete, curved
downward, without separate apophysis. There are isolated molars with
simple pointed crowns, seven or eight in number.

The sloths, living in trees and feeding on leaves that need to be crushed,
have jaws that are shorter and consequently stronger. The lower jaw is very
thick; its coronoid apophysis projects strongly; the part without teeth forms
a remarkable protuberance, above all in the unau or two-toed sloth—one
also sees this in the lower jaw of the elephant. The intermaxillary bone is
very small, which means that the maxillary also partly surrounds the opening of the nostrils—one scarcely sees this except in the rhinoceros, as a
result of the same small size of the intermaxillary bone. Finally, in the
sloths the zygomatic arch has at its base a fairly long descending apophysis, to which no quadruped shows any similarity (if one excepts the kanga-
roo or great jerboa of New Holland [Australia], the *Didelphis gigantea* of
Gmelin).[7]

If one now compares the head of our animal (see [fig. 7, bottom]) with
that of the sloths, one will find there all the characters minutely conserved,
despite the total difference that results from that of the proportion of the
parts. This apophysis of the notch of the zygomatic arch, this projection of
the anterior part of the lower jaw, the small size of the incisor bone [pre-
maxilla], and its distance from the nasal bones, are clear-cut characters
that leave no doubt.

The great thickness of the sides of the lower jaw, which even surpasses
that of the elephant, seems to indicate that the large animal we are examining doubtless did not content itself with leaves, but—like the elephant
and the rhinoceros—broke and crushed the branches themselves. The flat-
crowned serrated teeth would have been highly appropriate for this use.
The sloths have teeth that are more or less similar, but more separated.
Moreover, they have two more teeth in the upper jaw; but a still more
important difference is that their anterior teeth are longer, and pointed in
the form of fangs or canine teeth, which does not seem to have been the

7. See [fig. 7, top], the head of the unau or two-toed sloth; and [center,] that of the ai or three-
toed sloth.

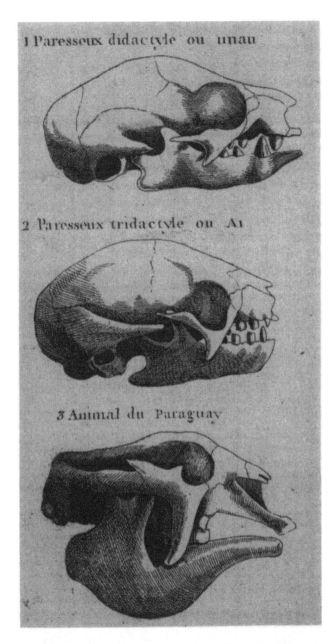

1 Paresseux didactyle ou unau

2 Paresseux tridactyle ou Aï

3 Animal du Paraguay

FIGURE 7 The skulls of the two-toed sloth *(top)* and three-toed sloth *(center)*, compared to that of the far larger fossil megatherium *(bottom)*; engravings illustrating Cuvier's paper. Note that all three skulls are drawn at the same size on paper, to facilitate comparison of their shapes and proportions.

case in the Paraguay animal. The position of the nasal bones in the latter have an affinity with that of the elephant and the tapir, which makes me suspect that it had a trunk; but it would have been very short, since the length of the neck and head together equals that of the forelimbs.

Be that as it may, we find in the absence of canine teeth, and in the shortness of the muzzle, characters sufficient to constitute a new genus in the family of edentates. It should be placed between the sloths and the armadillos, since it combines the shape of the head of the former with the dentition of the latter. It would be necessary to know particulars that this skeleton cannot give us, such as the nature of the integument, the form of the tongue, the position of the teats, etc., in order to determine more exactly which of these genera it approaches most closely. Meanwhile I believe I can give it the generic name of *Megatherium*, and the trivial name *Megatherium americanum*.

It adds to the numerous facts that tell us that the animals of the ancient world *[ancien monde]* all differ from those we see on earth today; for it is scarcely probable that, if this animal still existed, such a remarkable species could hitherto have escaped the researches of naturalists. At the same time it is a new and very powerful proof of the invariable laws of the subordination of characters, and of the justice of the consequences that have been deduced from them, for the classification of organisms *[corps organisés]*. In those two respects it is one of the most precious discoveries that have been made for a long time in natural history.

Translated from Cuvier, "Squelette trouvé au Paraguay" (Skeleton found in Paraguay, 1796).

4

A RESEARCH PROGRAM ON

FOSSIL BONES

In 1798, two years after his papers on elephants and the megatherium, Cuvier outlined what was now explicitly his own research agenda, in a paper to the Société d'Histoire Naturelle de Paris (Paris Natural History Society). A summary was published in the bulletin of the Société Philomathique, another informal scientific body in Paris, dominated by young savants such as Cuvier.

Cuvier explained that he planned to study the comparative anatomy of *all* fossil mammals, and he listed no fewer than twelve distinct species on which he had already started work. They included not only the mammoth and the megatherium, but also the puzzling "Ohio animal," fossil species of rhinoceros and hippopotamus, the huge-antlered deer or "elk" from the peat bogs of Ireland, an alleged bear from caves in Germany, a doglike carnivore from Paris itself, and several others less clearly defined. As in his paper on elephants (text 3), Cuvier concluded that it was not true that the species now living in the tropics had formerly lived at higher latitudes (as Buffon had argued); conversely, he claimed that these fossil species had had a wide geographical distribution but were truly extinct. What was new was Cuvier's final remark, clearly if covertly directed at self-styled "geologists" such as Faujas: "in view of this, it is up to geologists to make such changes or additions to their systems as they consider necessary to explain the facts that he [Cuvier] has here set out." The onus of explanation was shifted squarely onto the speculative devisers of geological "systems."

Soon afterward, Cuvier read another paper, this time to a meeting of the Institut open to the public, describing his research program in clear nontechnical terms (text 5). Significantly, he explained what was involved by using a vivid if unoriginal analogy. He was studying the "antiquities of nature," just as an antiquarian—or in modern terms an archeologist—studied the "monuments" of past civilizations; his research was providing reliable material for a *history* of the earth analogous to human history.

To make the topic attractive and immediate to a general audience, Cuvier chose as his main illustration the fossils that were being found in the gypsum quarries just outside Paris itself. The choice also had the major advantage that these fossils, like the megatherium but unlike most of the other fossil mammals, had scarcely been examined by any of his predecessors, and could be made his own intellectual property. Furthermore, since they came from strata of solid stone rather than loose superficial deposits, they posed much greater technical problems, and therefore displayed his practical skills to greater effect. Above all, however, these fossils were proving to be much more peculiar than even Cuvier himself had at first suspected: the Paris animal, he now claimed, was not a doglike carnivore after all, but instead filled a gap in the "échelle des êtres" (scale of beings) between the pachyderms and the ruminant mammals, or more precisely between the tapir and the camel. As an intermediate between two major groups, it was *more* unlike living mammals than any other fossil species: even the megatherium, by contrast, was unquestionably an edentate.

The example was a fine demonstration of Cuvier's zoological method. He claimed that it was his conception of the animal organism as a functionally integrated whole, constrained by its "conditions of existence," that had enabled him to assemble the skeleton from its disarticulated bones, without fear of constructing an imaginary monster from bits and pieces of different animals. He even sketched in words how, more conjecturally, one might go further, to reconstruct the whole animal body and infer its probable habits and habitat. He neatly forestalled any criticism of such conjectures, by saying he was being no *more* speculative than "geologists" were with their "systems"; and he even showed some sympathy with their ambition to account for the dramatic major features of the earth's history. But in the end, as before, he drew back to the conventional safety of concrete conclusions.

TEXT 5

Extract from a memoir on an animal of which the bones are found in the plaster stone [pierre à plâtre] around Paris, and which appears no longer to exist alive today.

Read at the public session of the National Institute on
15 Vendémiaire, Year VII [6 October 1798]

THERE IS NO LONGER anyone who does not know that the earth we inhabit shows everywhere clear traces of large and violent revolutions; but it has not yet been possible to unravel the history of these upheavals, despite the efforts of those who have collected and compared their documentation [documens].

The bones of quadrupeds found in the interior of the beds that form our continents are one of the most remarkable results of these revolutions. The thorough investigation of them that has been made in recent times has shown that they almost always come from animals alien to the climate in which they are found, or even from animals entirely unknown today. Henceforth it will therefore be necessary to add, to the [natural] history of the animals that exist at present in each country, that of animals that have lived or been transported there in the past. For this it will be necessary for physicists [physiciens][1] to do for the history of nature what antiquarians do for the history of the techniques and customs of peoples; the former will have to go and search among the ruins of the globe for the remains of organisms that lived at its surface, just as the latter dig in the ruins of cities in order to unearth the monuments of the taste, the genius, and the customs of the men who lived there. These antiquities of nature, if they may be so termed, will provide the physical history of the globe with monuments as useful and as reliable as ordinary antiquities provide for the political and moral history of nations.[2]

However, it is only with a rigorous and exact knowledge of comparative anatomy that one can proceed in this research without fear of error; it will

1. [The term "physics" still retained its older meaning, as a systematic study of the *causes* of any phenomena in the natural world (a meaning still preserved in the modern terms "physiology" and "physician"). So a "physicien" might be anyone who studied the causes of (say) electricity or digestion or mountains: in effect, anyone who might be eligible to belong to the First Class of the Institut!]

2. [The antiquarian metaphors were not Cuvier's invention. They had been a commonplace in the discussion of fossils since the late seventeenth century, and had figured prominently in Buffon's "Époques de la nature." But they may have been unfamiliar to Cuvier's audience, and in any case he was exploiting them with a new intensity of meaning.]

only be when the skeletons of all living species are thoroughly known that it will be possible to determine with certainty whether or not the bones that the earth conceals belong to some of those species. For as long as comparative anatomy was in its infancy, attention was given only to those bones that were striking because of their size or their unusual form, and they were regarded sometimes as the bones of giants, sometimes as those of elephants or other known species. But as this part of anatomy has been perfected, the study of them has been given greater precision; and Daubenton, Camper, and Pallas have been the first to contribute something fairly exact to this subject.[3] Today comparative anatomy has reached such a point of perfection that, after inspecting a single bone, one can often determine the class, and sometimes even the genus of the animal to which it belonged, above all if that bone belonged to the head or the limbs.[4]

This assertion will not seem at all astonishing if one recalls that in the living state all the bones are assembled in a kind of framework [charpente]; that the place occupied by each is easy to recognize; and that by the number and position of their articulating facets one can judge the number and direction of the bones that were attached to them. This is because the number, direction, and shape of the bones that compose each part of an animal's body are always in a necessary relation to all the other parts, in such a way that—up to a point—one can infer the whole from any one of them, and vice versa.

For example: if an animal's teeth are such as they must be, in order for it to nourish itself with flesh, we can be sure without further examination that the whole system of its digestive organs is appropriate for that kind of food; and that its whole skeleton and locomotive organs, and even its sense organs, are arranged in such a way as to make it skillful at pursuing and catching its prey. For these relations are the necessary conditions of existence of the animal; if things were not so, it would not be able to subsist.[5]

---

3. [Petrus Camper (1722–89) had been a distinguished anatomist in the Netherlands, and at one time professor of medicine at the University of Groningen. Peter Simon Pallas (1741–1811) was a Prussian naturalist attached to the Academy of Sciences in St. Petersburg; he traveled extensively in Siberia and other parts of the Russian Empire.]

4. [Note that Cuvier claimed to be able—in favorable cases—to "determine" or *identify* an animal from even a single bone; not, as legend suggested even in his lifetime, to *reconstruct* it. He did regard reconstruction as a legitimate goal, but only if most of the bones had been found (see below).]

5. ["Conditions of existence," along with "subordination of characters," are the key phrases that express Cuvier's conception of the living organism. Note the complete absence of the language of divine "design" in his description of the well-adapted organism: this is in striking contrast to the way most of his anglophone contemporaries—imbued with a tradition of natural theology—would have expressed the same point.]

I have chosen this example, as the most easily grasped and most appropriate, to give you an idea of the method that is employed in the research I am going to tell you about. You will readily appreciate that these kinds of relations between the parts are not all so obvious; and that as one descends to less important functions one is reduced to more subtle conjectures and to less firm conclusions; but at least it is always easy to assign to each of these results an appropriate degree of probability.

Among the bones I have examined following these principles, the most interesting and least known are those that are found in the plaster stone around Paris. This position, right inside the immense beds of gypsum that surround this city on the northern side, is in itself a remarkable circumstance. Most of the remains of quadrupeds that have been found hitherto occur in very loose deposits, such as the masses of sand or silt that could have been deposited by rivers, or indeed in the caves into which these animals could have retreated away from floods. Those I am talking about, by contrast, are embedded right in the interior of the stone, and must have been already scattered in the liquid in which it was formed and which took and enveloped them.[6]

Their consistency is very friable, and it is only by taking many precautions that they can be extracted. They are usually of a reddish tint. Their abundance is such that there is never a day when the laborers who work in the quarries at Montmartre, Mesnil-montant, Pantin, Argenteuil, and other nearby villages fail to find some in the blocks that they shape into building stones.[7] Vertebrae, ribs, and isolated teeth are the most common pieces; limb bones are rarer; and rarest of all are complete jaws and thin bones such as shoulder blades, because they are easier to break. Several connoisseurs [curieux] of this city have long collected the bones for their museums [cabinets], and it is by working through a large number of these collections that I have obtained the materials for this memoir [fig. 8].[8] The one that

6. [Cuvier refers to the medium from which the stone was deposited as a "liquid," because it was commonly assumed that it must have changed its chemical composition over time, as various rocks were successively precipitated out of it (see Cuvier's summary of Deluc's theory in text 2); only in the final stages of earth history did it become the ordinary saltwater of present seas.]

7. [These villages were outside Paris in Cuvier's day, but have since been absorbed by the growth of the city. Their quarries produced gypsum that was turned into the widely exported plaster of Paris; the associated stone was used for building.]

8. [The sheet reproduced as Fig. 8 is undated, but the description of the fossil as being from "l'animal moyen de Montmartre" (the medium[-sized] animal of Montmartre) implies that it came from an early phase of Cuvier's research: it must predate his recognition that two distinct genera were represented, but conversely it probably dates from somewhat later than the lecture reproduced here, at which point he seems to have considered that only one species was present.]

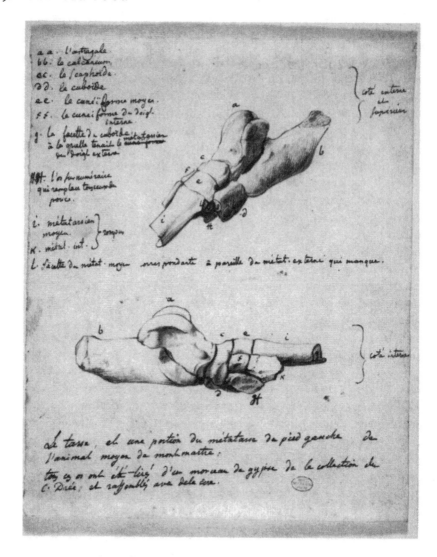

FIGURE 8 A sheet of Cuvier's working notes on specimens of fossil mammal bones from the quarries around Paris; the fine quality of his drawings is characteristic.

has provided me with the most was assembled by the late Mr. Joubert and now belongs to Citizen Drée, to whom I owe much acknowledgment for the generous way he has made them available to me.[9]

Having thus examined, described, drawn, and compared almost one hundred of these specimens, having matched them to one another according to

9. [Philippe Laurent de Joubert (1729–92) had been a legal official in Montpellier, and a keen fossil collector. After his death his important collection was acquired by Étienne Marie Gilbert, mar-

the clues given me by their articulatory facets, I have been able to reconstruct almost the entire skeleton of the animal to which they belonged. I
shall reserve for one of our closed sessions the details and the proofs of
all my work, which I shall accompany with the specimens on which it is
based.[10] Here I am only going to present the results, to give you an idea of
the skeleton of this animal, such as the specimens I have studied show it
must have been.

Its molar teeth have flat crowns, which show compartments of bony substance and of enamel. This is the structure of molars found in all animals
that feed on plants, because they have to have a kind of millstone for grinding, rather than the sort of scissors for cutting that carnivores have. The
particular form of these teeth is fairly similar to what one sees in the rhinoceros, that is, the upper ones are square and the lower are double cres-
cents; but the incisors are sharp, with six to each jaw, followed by a canine
on each side, behind which is a very short empty space. Without this presence of canines and incisors on the two jaws one would be tempted to take
our animal for a ruminant, because its molar teeth are so similar on the external face to those of the deer; but their crown is quite different.

This arrangement of its teeth is in general what one observes in pigs,
tapirs, hippopotamuses, rhinoceros, and other thick-skinned herbivores
with feet ending in several hoofs *[sabots]*. Thus, solely by an inspection of
these teeth, we can already judge that our animal belonged to the same
class. We shall see that all the rest of its skeleton confirms that conjecture.

The general form of its head, the curves and contours of its different
parts, have so much similarity to that of the tapir, that one is initially
tempted to regard it as coming from that South American animal. The
bones of the nose and muzzle are even formed in such a way that it appears
also to have had a short trunk like the tapir.

The front feet have three visible digits, the rear feet two. This is seen not
only by the facets on the wrist and instep, but also by specimens of stone in
which these feet have been found preserved whole. This number of digits is
all the more remarkable in that naturalists have not yet observed it in any
quadruped. It serves to complete the combinations that are possible in the
class to which our animal belongs: for the elephant has five in front and
five behind; the hippopotamus and the pig have four in front and four

quis de Drée (1760–1848), a Parisian politician, agronomist and naturalist, now demoted (tempor-
arily) to a mere "citizen."]

10. [At a "closed" session of the First Class of the Institut, only its members would normally be
present, so it was appropriate to go into more technical details. Note that "proof" *(preuve)* did not at
this time necessarily denote rigorous demonstration of a quasi-mathematical character; in both languages it was used roughly in the sense of "evidence."]

behind; the tapir four in front and three behind; and our animal has three in front and two behind. This places it immediately before the ruminants, which have two digits in front and two behind, and with which, as we shall see, it also has some similarities in other parts. Quite apart from its importance for the theory of the earth, the discovery of the animal thus serves to fill a gap in the scale of beings [échelle des êtres].

What I have just told you about the most important parts of its skeleton is sufficient to show that it differs essentially from all those that naturalists and travelers have hitherto discovered on the surface of the globe; and it is a proof of the great fact that I have already maintained in public,[11] that several animal species have been entirely destroyed by the revolutions that our planet has undergone. So I shall spare you a more detailed description of the other bones, which in any case could not be grasped without an actual inspection of the specimens.

The bones being well known, it would not be impossible to determine the forms of the muscles that were attached to them; for these forms necessarily depend on those of the bones and their ridges. The flesh being once reconstructed, it would be straightforward to draw them covered by skin, and one would thus have an image not only of the skeleton that still exists but of the entire animal as it existed in the past. One could even, with a little more boldness, guess [deviner] some of its habits; for the habits of any kind of animal depend on its organization, and if one knows the former one can deduce [conclure] the latter. After all, these conjectures would hardly be any more hazardous than those that geologists are going to find themselves obliged to make, in order to explain—within their systems—how the bones of an unknown animal come to be found in a country such as it is. And indeed, how can one fail to pardon some leaps of imagination, when warmed by so great a spectacle? How can one repress such a natural desire to give an account of causes that have been able to produce such terrible effects: to raise mountains, to shift seas, to destroy whole species, in a word to change the face of the globe and the nature of the beings that inhabit it?

But today only what can be observed or calculated is esteemed in the sciences; one cares little for what is [merely] guessed [se devine]. So I shall be content to have added some facts to the mass—already so impressive—that observers have collected, by showing (1) that the fossil bones that are found in the gypsum around Paris belong to an animal very different in form from all those that live in our climate today; (2) that this animal is not

---

11. [That is, in his papers on living and fossil elephants and on the megatherium, read at the Institut in 1796 (texts 3, 4).]

found alive in any country hitherto known; (3) that it forms a distinct genus which should be placed at the end of the pachyderm family, after the rhinoceros and the tapir, and immediately before the camel, which begins the ruminant class.

Translated from a manuscript, MS 628, Bibliothèque Centrale, Muséum National d'Histoire Naturelle, Paris. The original French text is transcribed in the appendix.

5

AN APPEAL FOR

INTERNATIONAL COLLABORATION

———

In 1800 Cuvier acquired two new positions, which he held in conjunction with his job at the Muséum. First, he was appointed one of the two secretaries of the scientific class at the Institut. Napoleon Bonaparte, who had made himself First Consul and virtual dictator by the coup d'état of Brumaire (November 1799), and who fancied himself a patron of all the sciences, chose soon afterward to take a turn as president of the Institut. Cuvier thereby came to know Napoleon personally, a contact that certainly helped his later career in governmental administration. Second, Cuvier was appointed to the prestigious position of professor of natural history at the Collège de France in succession to Daubenton. Not only had death now removed that senior colleague; Napoleon had earlier removed Cuvier's younger colleague Geoffroy, who had joined the team of savants that accompanied his military expedition to Egypt.[1] That left Cuvier for the time being in almost undisputed control of vertebrate zoology at the Muséum; although Faujas had also published work on fossil

---

1. Its most famous cultural prize was the Rosetta Stone, which later provided the key for deciphering the hieroglyphics of ancient Egypt (the stone was captured by the British while still in Egypt, and has been in the British Museum in London ever since). Geoffroy returned to Paris in 1801.

vertebrates, Cuvier had a low opinion of his senior colleague's competence in that or any other field.[2]

Cuvier too had been invited to go to Egypt; by declining, he had in effect chosen to build his career as a naturalist by working primarily in a museum rather than in the field. It was a shrewd decision, for he felt— as he put it much later—that he was already "at the center of the sciences." Anyway he was becoming well known among savants throughout Europe, not only for his major paper on elephants (text 3), now published in full, but also for an elementary textbook on comparative anatomy (*Tableau élémentaire*, 1798) and the first volume of his published lectures on the same subject (*Leçons d'anatomie comparée*, 1800–1805).

At this point Cuvier gave his research program on *fossil* anatomy much greater visibility. While borrowing some passages from his earlier lecture (text 5), his new paper for the Institut had much more ambitious goals. It was presented as a mere "extract" from a larger work in preparation; and it was addressed to "savants and amateurs of the sciences"—that is, to both experts and keen collectors—explicitly in order to enlist their collaboration (text 6).[3] As in his earlier paper, he introduced his subject with the commonplace idea that the earth has undergone major "revolutions" in the past. That term, in itself, merely denoted large-scale changes of state in the course of time, though his own inclination to infer sudden and violent events is clear enough. More significant is his sharp disjunction between the "ancient world" and the present: the one interpreted as in some sense chaotic, the other regarded by contrast as orderly.

Earlier attempts to create a history out of the apparent disorder and confusion of the ancient world, for example Buffon's, were dismissed as mere fantasies. They were contrasted with Cuvier's own heros, the naturalists of the late eighteenth century, who had rejected such speculative "systems" in favor of careful fieldwork: only a thorough understanding of the *present* world, Cuvier claimed, could lead to reliable knowledge of the past. He formulated the main problem in terms borrowed directly from Deluc: it was that of defining the event at the boundary between the present world and the past, the physical "catastrophe" that had resulted in

2. Cuvier is said to have punned unkindly on Faujas's name, calling him "Faujas sans fond" (Faujas without depth, substance, or content), which in French sounds similar to his real family name of Faujas de Saint-Fond.

3. Like "naturalist," the term "amateur" had no pejorative overtones of amateurism. It simply denoted someone who chose to pursue science—or, for that matter, music or literature or art—out of sheer love of the subject; the standard of knowledge and competence of such "amateurs" might be just as high as those who earned their living from the same activities.

the present form of the continents. Cuvier claimed that the large mammals that had lived before that event—the subject of his own research—could provide uniquely decisive evidence about it, since their living counterparts were more thoroughly known than any other group of animals. However, that required just the kind of careful anatomical comparisons that he himself was making, in order to establish that the fossil species really are different from the living. As before, he explained how such comparisons were necessarily based on his conception of the animal organism as a functionally integrated whole.

What was new in this paper was that for the first time Cuvier showed he was aware that his fossils were not all of the same age. He rejected the assertion that fossil bones were found only in loose superficial deposits *(couches meubles):* some came from solid strata. Those from near Paris even underlay beds containing the shells of marine mollusks, which implied that the animals had lived on land that had later been submerged before reemerging to become the present landmass. Based on decidedly scanty evidence, Cuvier even suggested that fossil animals differ from living species in proportion to their age.

What Cuvier presented as his most important conclusion, however, was his reiterated claim that *all* the fossils he could identify with confidence—his list was now almost doubled in length—were distinct from living forms, and therefore truly "lost" or extinct species *(espèces perdus).* For others his material was too fragmentary for positive identification; but Cuvier made it clear that he expected that they too would turn out to be distinct. The only exceptions he allowed were certain bones from peat bogs, but he dismissed these as not being true fossils.

Cuvier recognized that his conclusion left major unsolved problems: how had the fossil species been "destroyed" and how had their successors been "formed"? But as usual he declined to speculate on such matters. Specifically, he relegated the question of the origin of new species to "metaphysics," and thereby implicitly excluded it from the scientific realm of "physics," or the study of true causes.

The purpose of this paper, however, was not primarily to report on past research but to make claims for the future. Cuvier emphasized his thorough methods: his use of workmen in the local quarries to search for fossils, his inspection of private collections in Paris, and his correspondence with savants and collectors farther afield. His collaborators were carefully listed, in order to display his credentials; but above all, he openly appealed for more assistance of the same kind throughout Europe. Conversely, he tacitly discouraged any potential competitors by reporting how far his project was already advanced: particularly in the essential but expensive matter of illustration, he had already had no fewer than fifty plates of drawings engraved, ready for publication.

The scientific Class at the Institut set its seal of approval on Cuvier's project by ordering that the paper be printed as a separate booklet, so that it could be distributed without delay; doubtless Cuvier's position as one of the secretaries was an advantage here. Soon afterward the paper was also printed in the *Magasin encyclopédique* and in the *Journal de physique, de la chimie, et de l'histoire naturelle* (the periodical, now with a new name, that had been one of Cuvier's main sources of news when he was in Normandy); and extracts were published in German, Italian, and English journals. Cuvier's correspondence shows that this wide distribution was highly effective: information and drawings soon started arriving from all parts of Europe and even beyond.

TEXT 6

—

Extract from a work on the species of quadrupeds of which the bones have been found in the interior of the earth; addressed to savants and amateurs of the sciences, by G. Cuvier, member of the Institute, professor at the Collège de France and at the Panthéon central school, etc.

Printed by order of the Class for mathematical and physical sciences of the National Institute, on 26 Brumaire, Year IX [17 November 1800]

EVERYONE NOW KNOWS that the globe we live on displays almost everywhere the indisputable traces of vast revolutions: the varied products of living nature that embellish its surface are just covering debris that bears witness to the destruction of an earlier nature. Whether one digs into the plains, or penetrates into caves in the mountains, or climbs their torn flanks, one encounters everywhere the remains of organisms *[corps organisés]*, embedded in more or less thick beds that form the outer crust of the globe. Immense masses of shells are found at great distances from any sea, and at heights that it would be impossible for seas to reach today; beds of shale contain fish; seams of coal display the imprints of plants at heights or depths that are equally striking. But what is still more surprising is the disorder that reigns in the accumulation of these objects: here, shelly beds are covered by others that contain only plants; there, fish are superposed to terrestrial animals, and in turn have plants or shells above them. In other areas, lava flows and pumice stone, the products of subterranean fires, are mixed with products of the ocean. Almost everywhere these remains of organisms are utterly foreign to the climate of the ground that conceals them: it is in the tropics that one finds the living analogues of the fossil shells and fish of the north, and vice versa. In a word, just as nature has made the present habitats of living species attractive, and taken care to provide for their well-being and their conservation, so she seems to have been pleased to leave them with monuments of her power in this disorder and apparent confusion, and clear proofs of the upheavals [bouleversements] that must have preceded the present order of the universe.[4]

These traces of devastation have always been striking to the human mind. The legends [traditions] of deluges that are preserved among almost all peoples are due to the marine fossils [corps marins] scattered over the whole earth. Legends of giants—no less universal—derive from bones that are larger than those of any of the [living] animals of the climates in which they are found from time to time. But these are not just vulgar ideas. Men of another kind have sought to comprehend the whole generality of the phenomenon, in order to ascend to its causes. They have excavated in the ruins of the globe in order to find monuments of its physical history, just as antiquarians excavate in the ruins of cities in order to find monuments of the history of the crafts [arts] and customs of the people who lived there. The Woodwards, Whistons, Leibnizes, and Buffons were unable to contemplate these objects without the disquiet that is a mark of genius.[5] Their imagination, fired by such a grand spectacle, shot back into the past, and believed itself present at these successive catastrophes, inundations, subsidences, and conflagrations; they believed they were tracing a history, when it was only that of their own creation.

Like men, however, the sciences have their stages of life. Given up in youth to brilliant imaginative illusions, they become cooler and more reasoned in maturity. The creative geniuses that give them birth thrust them into a career by a kind of inspiration, and they follow it almost recklessly; and it is necessarily so. Timid spirits would start by noticing the obstacles; but daring minds surmount them without perceiving them, and their example encourages the timid. They in turn become involved; their progress is slower, and they take no step without having recognized its difficulties

4. [This personification of "nature"—taking the place, in effect, of a providentially wise Creator—was a common feature of scientific writing in the Enlightenment; the effect is heightened in French by the feminine gender of the word "nature."]

5. [Buffon is here implicitly dismissed, by being associated with the authors of still earlier and—in Cuvier's opinion—even more speculative "theories of the earth."]

and smoothed it out. The first guessed at nature rather than studying it; the others, while thinking they are only verifying the systems they admire, study it truly; and it is thus that the sciences—like peoples—pass from poetry to history.

The theory of the earth has thus taken a new direction in the past twenty years. The Saussures, Pallases, and Dolomieus were less eager to attract the admiration of their contemporaries by brilliant but fragile edifices, than to set in place some solid foundations on which posterity could one day construct a lasting monument.[6] They rejected all "system"; they recognized that the first step to make in divining the past was to establish the present firmly. Since then, instead of imagining causes, one has collected facts. Mountains, veins, and strata have been penetrated in all directions; one has assembled their materials and compared them with one another; and already we possess a mass of genuine knowledge that far surpasses all that could have been hoped for when this method began to find favor.[7]

There is, however, one part of the animal kingdom whose fossil remains have been less studied, namely the quadrupeds. Attention has long been given only to those fossil bones that are striking for their size or unusual form. Sloane, Messer-Schmidt, Daubenton, and Pallas have thus made us aware of the bones of elephants and rhinoceros scattered in northern countries, and have given birth to the quite widespread idea that tropical animals formerly lived in the north or that they were swept there by some inundation that took that direction.[8]

Continuing this research, Camper, Blumenbach, Hunter, Rosenmüller, and Faujas have well realized that such a cause is insufficient to explain all the phenomena, and that the distribution of the fossil bones is nothing like as regular as had been imagined; but they have not exhausted the matter.[9]

6. [Like Pallas, Saussure and Dolomieu had both been highly active in geological fieldwork. Horace-Bénédict de Saussure (1740–99) had spent his life based in his native Geneva, but had explored the Alps intensively and traveled widely in other parts of Europe; he had died only the year before Cuvier's address. Déodat Guy Silvain Tancrède de Gratet de Dolomieu (1750–1801) had taught at the École des Mines in Paris, and had done much fieldwork in France, before joining the Egyptian expedition. In 1799, on his return journey, he had been taken prisoner of war; he was released a few months after Cuvier delivered this paper, and returned to Paris as Daubenton's successor at the Muséum, but he died soon afterward.]

7. [In this passage the French pronouns have been retained in translation, although this makes the English somewhat stilted: they show how Cuvier slides deftly from "they" (the older naturalists) through the inclusive "one" to the "we" that clearly embraces himself.]

8. [Sir Hans Sloane (1660–1753) had been a famous English naturalist whose vast and varied collections had become, after his death, the core of the new British Museum in London. Daniel Gottlieb Messerschmidt (1685–1735) had been a Prussian naturalist who traveled in Siberia.]

9. [Johann Friedrich Blumenbach (1752–1840) was professor of medicine at the University of Göttingen and a distinguished anatomist and physical anthropologist. William Hunter (1718–83) had been a fashionable and wealthy surgeon in London and had contributed to earlier debates on

Comparing the number of species they have examined with those that still remain, it could almost be said that they have scarcely skimmed it.

However, this kind of fossil has no less interest than others for the theory of the earth; one could even say that it is easier to reach a decisive result by examining the bones of quadrupeds than by studying those of all other animal fossils. The principal question being to know the extent of the catastrophe that preceded the formation of our continents,[10] it is above all a matter of determining whether the species that then existed have been entirely destroyed, or solely modified in form, or simply transported from one climate to another.[11] Now it is clear that it should be easier to make such an examination in the case of the mammal class than of all the others: it is the least numerous, and we [already] know almost all its species. If there are still some to be discovered, they are surely small and unimportant. It is almost impossible that any of the large ones have escaped pursuit by travelers and the inquiries of naturalists. Furthermore, we now possess the skeletons of almost all that are known. We can thus compare, and decide with sufficient certainty whether any fossil bone does or does not resemble the analogous bone[12] in living species. The case is not the same with shells and fish: naturalists are still far from having observed all of them; and each time we find an unknown [fossil] fish or shell in the earth we can [always] suppose that the species is still living in distant seas or at inaccessible depths.

Despite these reasons for preferring the study of the fossil bones of quadrupeds, the distinguished men I named above have been hampered in their research by two kinds of difficulty. First, these bones are more difficult to collect than all other fossils, and are rarely found well preserved. The workmen who come across them give them little attention, because they take them for the bones of men or ordinary animals; often even savants have not perceived the subtle differences that distinguish them from those of common species. Second, it is not easy to establish the neces-

---

fossil bones. Johann Christian Rosenmüller (1771–1820)—a near contemporary of Cuvier—was professor of anatomy and surgery at the University of Leipzig.]

10. [A formulation borrowed unmistakably and unquestioningly from Deluc. Cuvier is sure to have read Deluc's most recent exposition of his ideas, in his *Lettres à Blumenbach* (1798).]

11. [Cuvier's formulation indicates how extinction, transmutation (evolution), and "transport" (whether by migration or by a violent flood) were regarded as three *alternative* explanations for the *same* problem, namely the lack of identity between fossil and living species.]

12. [Like all naturalists at this time, Cuvier uses the term "analogous" *(analogue)* to mean what was later distinguished as "homologous"; that is (in this case), for bones that have the equivalent place in the skeleton, whether or not they had the same function and were "analogues" in the modern scientific sense.]

sary comparisons everywhere. It is virtually just in our own day that comparative anatomy has emerged from infancy, and there are hardly two or three places in Europe where the collections are complete enough to contain all the specimens necessary for precise comparison. It is to these two causes that we should attribute the imperfection of our knowledge of the subject in question, and the errors that dominate [even] the most estimable works.

I have already mentioned above the error that only the bones of tropical animals are found in the north. Several authors still think that these bones are exactly the same as those of living species; that they are never found except in loose deposits [terrains meubles], where they could have been transported by rivers; that South America has no fossils of animals of the tropical zone of the Old World, although there are some in North America; and that the Old World has no bones that belong to animals of the New. It is because the phenomenon has thus been poorly identified, that it has been thought possible to explain it by the suppositions of a perpetual spring, an inclination of the axis of the globe, a displacement of ocean basins, a gradual cooling of the earth, and still others equally inadequate.

Having acquainted myself with the causes of these inaccuracies, I thought I should concern myself with destroying them. I started by reviewing all that had been done on this subject by my predecessors. I have compared afresh the bones they mentioned, which I was able to procure, with their living analogues. I have employed men to search for the bones that are concealed in the quarries in the vicinity [of Paris]; I have visited the cabinets in which they had been deposited; I began correspondence with different countries, and the savants who live in them have sent me descriptions and drawings of the fossil bones they have discovered there. I should say that I have been supported with the most ardent enthusiasm and the most noble disinterest, not only by my friends, but also by all the Frenchmen and foreigners who cultivate or love the sciences, whom it has been possible for me to interrogate. The result of this combined effort has been the most complete report that has yet been assembled, on the various bones that have hitherto been recovered from the bowels of the earth.

But this was not sufficient. It was necessary to determine the genus and species of each bone, of each substantial fragment of bone. It was necessary to assemble the bones belonging to the same species, to reconstruct in some way the skeletons of the animals; and then to compare the beings thus revived [ressuscités] with those that naturalists have discovered alive on the surface of our present earth, to determine their similarities and their differences. I will say more: it was necessary to penetrate within these frameworks *[charpentes]* to their real nature and to the way of life of the animals from which they came.

This claim is not at all as fanciful *[romanesque]* as it will perhaps appear to those who have no idea of the method that is followed in this kind of research.

In the living state, all the bones are attached to each other, and form an ensemble among which all the parts are coordinated. The place that each occupies is always easy to recognize by its general form, and by the number and position of their articulating facets one can judge the number and direction of those that were attached to it. Now the number, direction, and shape of the bones composing each part of the body determine the movements that that part can make, and consequently the functions it can fulfill. Each part in turn is in a necessary relation with all the others, such that up to a certain point one can infer the ensemble from any one of them, and vice versa.

For example, when the teeth of an animal are such as they must be, for the animal to feed on flesh, we can be sure without further examination that the whole system of its digestive organs is adapted for this kind of food, and that its whole framework, its organs of locomotion, and even its sense organs, are made in such a way as to make it skillful in perceiving, pursuing, and seizing its prey. In effect, these relations are the necessary conditions of existence of the animal, and it is evident that if things were not so this animal could not subsist.

I have chosen this example as the most palpable and the most appropriate to make conceivable the kind of reasoning that this research demands. It can easily be sensed that not all the relations of the parts are as demonstrable as these, and that one is often reduced to more tentative conjectures and less certain conclusions; but at least it is always easy to assign the degree of probability that belongs to each of these results. Besides, one does not always have to work with isolated bones. Very often it happens that almost complete limbs are discovered; sometimes no part of the skeleton has been separated from the others. In these happy cases the anatomist has almost nothing to do, for—I repeat—the skeleton determines the form of the soft parts; and imagining those in turn covered by the skin, one has the animal as it was when alive, apart from features of almost no importance such as crests, manes, and other purely external parts with no influence on its inner nature.

It is by studying the fossil bones of quadrupeds on these principles, that I have reached the results that I am going to expound in a general way, and for which I shall give the evidence, with all the inferences *[développemens]* that can be drawn from them, in the work for which the present memoir is as it were the prospectus [programme].

First, one finds abundantly, under the soil in all countries, bones different from those of the animals that live at the earth's surface today.

I say *abundantly*, for in all localities where a little care has been taken to look for them, a great number have been found. There is not a day, for example, when the laborers who work in the plaster quarries around Paris do not discover some; and if there are not more in collections, it is because collectors [curieux] have not shown enough interest in them, and the workmen have thrown them away, failing to recognize their value.

I say *in all countries*, because it is only those that naturalists have not yet been able to study at leisure, that have not yielded any. The soil of Siberia swarms with them. There is hardly a region of Germany, Italy, France, England, Ireland, or Spain that does not have some. Ever since America has been examined by educated people living locally, it too has yielded them. Those from the banks of the Ohio have long been known; Dombey has found others in Peru.[13] The Spaniards have reported a complete skeleton from Paraguay. The Philosophical Society of Philadelphia has just made some more known from the United States.[14] Tartary [central Asia] has yielded some of them; and although we do not yet have any of them either from Africa[15] or from the large continent of New Holland [Australia], there is every reason to believe that that is simply owing to lack of research.

Finally, I said that *these fossil bones are almost always different from those of the animals that live on the ground that conceals them*, even when otherwise they have a more or less complete resemblance to those of animals of other countries. Stony or earthy beds have no longer been forming on our continents, since they have enjoyed their natural [i.e. present] climate. That is, when animals have died, their bones, exposed to all the effects of the atmosphere, are not slow to decompose. Decomposition is fairly slow, although no less real, when these bones are buried in a loose deposit [terre meuble], as happens in our cemeteries and drainage ditches. Only stony stalactite is able, by enveloping them, to preserve them from corruption;[16] apart from that, it is almost impossible today for fossil bones to be

13. [Joseph Dombey (1742–94) was a French naturalist who had traveled widely in South America.]

14. [Probably an allusion to Jefferson, "Bones of a quadruped" (1799), published in that society's *Transactions* the previous year.]

15. It is said that there are some at Ceuta absolutely similar to those of Gibraltar. [Ceuta is on the Moroccan coast opposite Gibraltar.]

16. ["Stalactite" was used to denote any stony incrustation, not just the icicle-like forms found in some caves.]

formed, and in effect we find none at all newly formed. Nowhere are there any human bones: all that has been said to contradict that assertion has been found false, whenever it has been possible in good faith to examine the bones that were claimed as such.

Some authors, most recently Mr. Deluc, have thought that the fossil bones of quadrupeds are always found in loose deposits *[couches meubles]*, the most recent of all those that envelop the core of the earth.[17] This is not generally so. Often they are embedded in true stone, either calcareous or gypseous or even siliceous; and not only in caves or in fissures in the rock, where — as I have just said — stalactite could have enveloped them recently, but also in the natural beds of these rocks, and sometimes of very ancient rocks. In this way those around Paris are in the middle of enormous beds *[bancs]* of plaster, covered in turn by beds of oysters and other marine shells. I even believe I have noticed a fact still more important, that has its analogies in relation to other fossils: namely that the older the beds in which these bones are found, the more they differ from those of animals that we know today.

But it is the generality of this difference that makes it the most remarkable and astonishing result that I have obtained from my research. I can now almost assert that none of the truly fossil quadrupeds that I have been able to compare precisely has been found to be similar to any of those alive today.

I am well aware that if it were only a matter of the testimony of authors, and even of respectable authors, one would find much to oppose me. Without mentioning the naturalists of yore, who found "human" fossil bones everywhere, in our own day Gouan and Spallanzani say they have found them; Esper claims that the bones in the caves in Franconia are the true bones of polar bears; Pallas, that the mammoth of Siberia is in every way similar to the elephant; and so on with others.[18]

But this testimony soon evaporates under scrupulous examination; and when some doubt remains, it is because the bones being examined are such that they differ no more than from one living species to another living species. For example, all the ruminants have teeth so similar that they can be distinguished only by their size: thus two species of the same size have teeth that are absolutely alike. It is thus impossible to conclude, from the identity of the tooth of a fossil ruminant with that of a living species, that it did not come from a different animal. Apart from this single case, all the

17. [When Deluc read this, he wrote to Cuvier to deny that he had ever made such a claim.]

18. [Antoine Gouan (1723–1821) was a physician and botanist in Montpellier. Lazzaro Spallanzani (1729–99) had been professor of natural history at the University of Pavía. Johann Friedrich Esper (1742–1810) was professor of natural history at the University of Erlangen.]

complete fossil bones that I have seen are different from those of living quadrupeds.

After lengthy research, and with the help of my predecessors and friends, I have been able to restore [rétablir] twenty-three species, all quite certainly unknown today, and which all appear to have been destroyed, but whose existence in remote centuries is attested by their remains.[19]

The one discovered longest ago is that of which the tusks yield the fossil ivory so common in Siberia [i.e. the mammoth]. It was agreed that it could be regarded as the same as the Indian elephant; but I have shown in another memoir [text 3] that it differs quite substantially, and it was known before me that it usually surpassed it in size. Its remains are found all over Europe and Asia, right to the shores of the Arctic Ocean [mer glaciale]. Another almost complete skeleton of it was found last year near Gotha, in the same area in which another was found at the beginning of this [i.e. the eighteenth] century. A valley in the region of Canstatt in Swabia has furnished eight skeletons of it. Two years ago a considerable part of one was found near the village of Argenteuil, two leagues [6 miles] from Paris. It would be impossible to detail here all the places where it has been unearthed.

The second of these species is that to which the English and the inhabitants of the United States have transferred the name of *mammoth*, which properly belongs to the first. It is as large as the previous one, but its enormous teeth, armed with points, give it a distinctive character.[20] A huge quantity of its bones is found in an area on the banks of the Ohio River, in the west of the United States; almost all those in collections in Europe and America are from there, but this species is also found in Siberia, Little Tartary [European Russia], and Italy.

The third lost species is that of the long-headed rhinoceros, which, as I have shown in another memoir, is essentially different from the four or five species or varieties of living rhinoceros; it is common in Siberia and Germany. A complete one has been found embedded, with its skin and flesh, in the frozen land beside the Vilhoui, a river that flows into the Lena; which—to mention it in passing—proves that the revolution that destroyed the animals I am speaking about was extremely sudden.

The fourth lost species is that which, in a separate paper [text 4], I have named megatherium; this resembles on a large scale the quadrupeds called sloths. A complete skeleton was found in Paraguay, and is now conserved in

19. [The low antiquity implied in this clause is in striking contrast to Cuvier's casual assumption of "thousands of centuries" a few years later (see text 8); but here it applied primarily to the bones in the superficial deposits, whereas the latter phrase would refer to those in the Parisian gypsum formation, which, at least by that time, Cuvier realized is far older.]

20. [Cuvier later named it *Mastodon.*]

the museum of the king of Spain; a very fine description of it has been published in Madrid. Its remains are also found in North America, for the megalonyx described by Mr. Jefferson does not seem to differ from it at all.[21]

The fifth species is the large bear of which the bones are present in enormous quantities in some caves in Germany; and which has been recognized by [Petrus] Camper and Rosenmüller as very different from living bears. Another species of bear, which is found mixed with the preceding one in the same caves, will form my sixth species; [Adriaan] Camper the younger and I have been the first to recognize its differences.[22] A species of carnivorous animal from the same caves, intermediate between the wolf and the hyena, will form the seventh.

The eighth species will be the animal related to the elk, which is found in such abundance in Ireland, and of which the antlers are up to fourteen feet across from one tip to the other. The English have described its bones several times.

The ninth will comprise the large fossil turtles found in several countries, which it seems should be divided into several species.

The tenth is the large animal that passes for being of the lizard genus and which is so well known under the name of the Maastricht crocodile. The Campers (father and son) have devoted much study to it, and Citizen Faujas has just given a complete description of it, as well as of the quarries in which its bones are found.[23]

The eleventh will be the very remarkable reptile embedded in the shales around Eichstatt, of which Mr. Collini has described an almost complete skeleton, conserved in the museum at Mannheim. It was small, and appears to have enjoyed the ability to fly, as the little lizard called the "dragon" does today.[24]

The twelfth is another animal, either reptile or whale, also described by Mr. Collini.

Apart from these twelve species, the bones of which have been discovered or identified by others, I have collected or been the first to recognize the characters of eleven others, most of which are found in France; namely:

21. [Cuvier later changed his mind on this. Jefferson had just published his paper ("Bones of a quadruped," 1799) in Philadelphia, based on fossils found in Virginia, and had named the animal "megalonyx" (great claw).]

22. [Adriaan Gilles Camper (1759–1820) had inherited his father's fine collection, and his interest in anatomy; he had first made contact with Cuvier (by correspondence) only the previous year.]

23. [Cuvier later named it *Mosasaurus*.]

24. [Cuvier later named it "pterodactyle" (wing-fingered); it was in modern terms the first pterosaur to be discovered, and came from the famous lithographic stone at Solnhofen in Bavaria. Cosimo Alessandro Collini (1727–1806), a native of Florence, was director of the natural history museum in Mannheim; he had earlier been Voltaire's secretary in Berlin.]

1. The animal whose teeth, impregnated with copper, yield the western turquoises. Many are found at Simore in Languedoc, where there was once a quarry for these turquoises. One of its teeth has also been found near Trévoux. Dombey has reported from Peru some teeth that appear to be of the same species, many of which are impregnated in various places with native silver. This species is very close to that of Ohio.

2. A species of tapir of which the bones are also found in Languedoc, on the slopes of the Montagne Noire; it is of the same size as the living tapir, which (as is well known) comes from South America, and differs only in the form of its last molar teeth.

3. A second species of tapir, which I call giant on account of its size, which equals that of an elephant; but its form does not differ at all from that of the ordinary tapir. Its remains have been found near Comminge and near Vienne in Dauphiné.

4. A species of hippopotamus, which resembles in miniature the living hippopotamus, and which is no bigger than a pig. I have found its bones in a siliceous sandstone of unknown provenance.

5–10. The plaster quarries around Paris alone have given me six fossil species, three of which I have already spoken about elsewhere. All six are of a genus hitherto unknown, intermediate between the rhinoceros and the tapir. The differences between them consist above all in the number of digits in the feet, and in their size, which ranges from that of a horse to that of a rabbit. I have such a large number of the bones of these species that I could reconstruct [rétablir] their skeletons almost completely.

11. Finally I have just recently discovered the existence near Honfleur of the bones of a species of crocodile, very close to that called the gavial or crocodile from the Ganges, but nonetheless easily distinguished from it by some striking characters.

So there, indeed, are the twenty-three species of animals unknown to-day that I was certain of possessing. But those concealed in the earth are not limited to these; and the following data—which I have not wanted to put on the same level as the preceding, because they do not have the same degree of certainty—are nonetheless sufficient to make us hope that we shall soon be able to enlarge this catalogue of zoological antiquities.

I arrange these still uncertain data in three classes. First, I know some fossil specimens fairly similar to the equivalents in living species, but which come perhaps from species that differ in other parts. Such are

1. The bones of quadrupeds of the tiger genus, mixed with those of the bear that I spoke of above. The specimens I have seen show hardly any difference from their analogues in the tiger and lion.

2. The head of a hyena, described by Collini and regarded by him as that

of a seal. To judge from his drawing and description, it differs in no way from that of an ordinary hyena.

3. The bones from the rocks of Dalmatia. I have seen some teeth from there that are exactly like those of the fallow deer [daim], but perhaps the animal differed in its antlers.

Next I have seen some other specimens that are not complete enough to recognize clearly their identity or nonidentity with their [living] analogues. Such are

1. The bones of large ruminants from the region of Verona.

2. Those of the same class from the Rock of Gibraltar.

3. The bones of rodents from the same Rock.

4. The bones of cetaceans of the dolphin genus or that of the sperm whale, which Mr. Deborda d'Aureau found near Dax, and which he thought belonged to crocodiles.

5. The bones of ruminants of several different sizes, some of which are like sheep, found at Mont Abuzard near Orléans.

6. The bones from around Aix, from Cette [Sète], etc. I have indeed seen some specimens of them, but so badly damaged that I cannot even identify their class.

7. I have also heard or read accounts of a multitude of places where bones are said to be found, but of which I have seen none. Such are the various caves in the Crapac mountains, the Harz, the Dalmatian islands, the isle of Cerigo [Kythera], the environs of Concud in Aragon, those of Cadiz, etc.

Finally, in the third class of uncertain bones I put those that are completely like living species, but which, having been found in peat bogs, could have been buried there by various causes, without having to be regarded for that reason as true fossils. Such in particular are the bones of cattle, buffalo, aurochs, and water buffalo, which are so frequent in the marshes and peaty depths of Europe and Asia. Siberia, Germany, Holland, Scotland, and above all the Somme valley in France have all yielded a large number of them. Here then again are several unidentified species, some of which will probably need to be added to the twenty-three that have been identified.

This remarkable number has been collected or identified in only two years, and that by a man who has utilized no other means than his own zeal and the favor of several friends of the sciences. From that can be judged what the attention of naturalists could produce, aroused by these first findings and, above all, in due time, that element so necessary for the perfection of all our knowledge. If so many lost species have been restored [rétablies] in so little time, how many must be supposed to exist still in the depths of the earth! How much will the ideas we already had about the revolutions of the globe be enlarged by these circumstances that were hitherto unknown: animals that formerly lived on the earth's surface, buried under entire mountains; between them and the present surface, traces of the successive passages of seas; an earth, a primitive nature, which was not at all submissive to the empire of mankind, and of which only some half-decomposed bones remain to us! How were these antique organisms [êtres antiques] destroyed? Is not metaphysics itself even more embarrassed by these facts than simple physics? And is not this new production of organisms perhaps more inconceivable than any other part of the phenomenon?

It seems to me at least that what we have already recognized is important enough to commit us to further research, and I hope the friends of the sciences will want to continue to favor me. I only ask them for what it is impossible to obtain without their friendship: I mean reports of fossil bones in their possession or at their disposal. If they are willing to let me have drawings made of these bones, I will defray all the costs that those drawings entail. For my part, I shall endeavor to render them all the services that are in my power, by identifying the objects that I have at my disposal to study, and which could be useful in their own study and research. This reciprocal exchange of information [lumières] is perhaps the most noble and interesting commerce that men can have. I shall take the greatest care to record in my work the names of all those who will have contributed to its perfection, and I shall make use of the discoveries that are communicated to me, only in assigning glory to their true authors.

The most celebrated foreign naturalists, Messrs. Blumenbach, Camper, Fortis, Fabbroni, Brugmans, Autenrieth, Jäger, and Wiedemann;[25] my colleagues Lacépède, Faujas, Daubenton, Hermann, Gillet, Lelièvre, Bosc, Brongniart, Dolomieu, and Fischer;[26] the owners of the finest collections,

25. [Giovanni Battista (Alberto) Fortis (1741–1803), a priest in the Augustinian order, was a naturalist in Bologna, well known for his extensive travels. Giovanni Valentino Mattia Fabbroni (1752–1822) was a naturalist who held various official positions in his native Florence. Sebald Justin Brugmans (1763–1819) was professor of medicine and chemistry at the University of Leiden. Johann Hermann Ferdinand Autenrieth (1772–1835) was professor of medicine at the University of Tübingen. Karl Christoph Friedrich von Jäger (1773–1828) was a physician and naturalist in Stuttgart.]

26. [Bernard Germain Étienne de la Ville-sur-Illon, comte de Lacépède (1756–1825), was professor of natural history at the Muséum and a prominent French politician in Paris. Johann Hermann (1738–1800) was professor of the natural and medical sciences at the University of Strasbourg. François-Pierre-Nicolas Gillet de Laumont (1747–1834), Claude Hugues Lelièvre (or Le Lièvre, 1752–1835), and Alexandre Brongniart (1770–1847) were all in the Corps des Mines in Paris. Louis Augustin Guillaume Bosc (1759–1828) was an agriculturalist and horticulturalist in Paris. Gotthelf Fischer von Waldheim (1771–1853) was professor of natural history at the University of Mainz (and later held a similar position in Moscow); he too counted as a "colleague," because Mainz had been annexed by France, and its university brought within the French system.]

Drée, Besson, and Saint-Genis;[27] the trustees of several public museums in France and abroad: [all] have helped me with their advice and with facts that have come to their attention, and have informed me about the specimens that are found at their disposal.

Such men should encourage others to follow their example, and I have no doubt that they will find worthy imitators. It is with this confidence that I have requested the class of the Institute to which I have the honor to belong, to recommend me in some way to men who could be useful in my enterprise, by ordering the printing of the prospectus of my work. The favor it has shown me, in acceding to my request, is a sure guarantee of the welcome I shall have from the savants of Europe. Besides, I believe I have a kind of right to that welcome, by the highly advanced state of my work. Already I have more than three hundred drawings; fifty plates have been completely engraved and many others started; and I am waiting for nothing more, before having my book published, than the information that the present paper can procure for me.

The Botanic Garden at Paris
10 Frimaire, Year IX [1 December 1800]
G. Cuvier

Translated from Cuvier, "Espèces de quadrupèdes" (1801).

27. [Auguste Nicholas de Saint-Genis (1741–1808) was a lawyer, writer, naturalist, and agronomist in Paris. The other reference may be to Alexandre Besson (1758–1826), a former Jacobin politician who at this time was director of the saltworks in the east of France.]

6

THE ANIMALS FROM THE GYPSUM

BEDS AROUND PARIS

————

The Peace of Amiens in 1802 made England accessible to Frenchmen for the first time for many years, and Cuvier planned a visit to London to add the rich English collections to his store of material on fossil bones. But that plan was aborted by the first of his many governmental appointments: he traveled for several months in the south of France, directing the reorganization of secondary education there.[1] Just before he left Paris he was appointed full *(titulaire)* professor at the Muséum, on the death of Mertrud; he took the opportunity to have the chair redefined as "*comparative* anatomy." While he was away, his function as one of the secretaries of the scientific Class at the Institut was converted into a highly paid tenured position *(secrétaire perpétuel)*. As he punned to an Italian friend, its duties now made him "perpetually fixed" in Paris;[2] but it did finally establish the financial and professional security of his career. The following year, at the age of thirty-four, he married the widow of a victim of the Revolution—like Lavoisier, a tax

1. By the time Cuvier returned to Paris, the fragile peace had collapsed, and the renewed war delayed his first visit to England until 1818, after the fall of Napoleon.

2. Quoted in Outram, *Georges Cuvier* (1984), p. 67, from a letter to Fabbroni in Florence.

collector *(fermier-général)*—and acquired a ready-made family of four children.[3]

Meanwhile the foundation of the *Annales du Muséum* had given Cuvier and his colleagues their own medium of publication, including— what was indispensable for the natural history sciences—a generous allowance of engraved illustrations. In the first volume (1802) Cuvier and Geoffroy, together with their senior colleague Jean-Baptiste de Lamarck (1744–1829), the professor of *insectes et vers* (i.e. invertebrate zoology), reported jointly that the mummified animals that the Egyptian expedition had found in ancient tombs were all of modern species. The possible significance of that fact in relation to theories of transformism (or evolution) received no comment, probably because the three authors could not agree on what it was (see text 19, sec. 33). Cuvier also contributed several papers on molluscan anatomy. The next volume carried a formal report of the establishment at the Muséum of Cuvier's separate collection for comparative anatomy; four papers by Faujas, three of them on fossil vertebrates, showed that Cuvier still had no monopoly on such material. In the third volume (1804), however, Cuvier began to publish what soon became an astonishing torrent of papers on the bones of fossil animals. These were in effect preprints of what, eight years later, he collected and reissued as his *Recherches sur les ossemens fossiles* (Researches on fossil bones, 1812).

In fact the first two of these papers were not on fossils at all. If the osteology of a relevant living species was not well known, Cuvier presented a study of it as a prelude to his subsequent analysis of the related fossil species, so that the basis for the comparison was fully in the public realm: nothing could have indicated more clearly how his method was to base inferences about the past on knowledge of the present. One of the first papers, for example, dealt with the poorly known living tapir, and was followed immediately by a paper in which some fossil bones and teeth were identified as tapir-like. It is no accident that Cuvier chose this particular paper to make his debut as a fossil anatomist in the *Annales*. Its conclusion was striking, because the living animal was known only from South America, whereas the fossils were from France. He made sure the moral was clear to geological theorists. He claimed that "all hypotheses based on the Asiatic origin of our [European] fossils are hereby

---

3. Cuvier's subsequent family life was marked by tragedy. His own first child died in infancy, and two others—including the next Georges—died in childhood; such mortality was all too common at the time. In 1827 his last child, Clémentine, died at the age of twenty-two, shortly before her marriage; Cuvier found the loss almost unbearable, and it clouded the last five years of his own life.

destroyed"; and he added, in what was becoming a familiar refrain, that "geology" needed the "touchstone" *(pierre de touche)* of reliable facts on which to build its "systems."

Other papers published by Cuvier in the *Annales* the same year included a full description of the megatherium, with a translation of what Bru had published in Spanish soon after Cuvier's report (text 4) first upstaged him. Cuvier also analyzed the fossil animal that Thomas Jefferson had first reported from Virginia and named the megalonyx: Cuvier now interpreted it as a cow-sized carnivorous sloth that had probably preyed on the "Ohio animal." In addition to such papers on fossils, Cuvier also threw in half a dozen on living mollusks, for good measure.

Cuvier's most important papers, however, concerned the fossils from around Paris. As already mentioned, these came as it were from his own and his colleagues' doorstep. Although they had been keenly sought by collectors, they had been little studied by Cuvier's seniors in the field, and they were a much tougher challenge for his methods than any of the more recent fossils. A long series of papers, stretching over the next four years, dealt with successive parts of the skeleton of the commonest fossils, starting with the skull and teeth as the most revealing parts. Since his lecture on them at the Institut six years earlier (text 5), further study— and not least the discovery of significant new specimens—had in fact convinced Cuvier that they belonged to two separate genera with several species in each. He named the genera *Palaeotherium* (ancient beast) and *Anoplotherium* (unarmed beast). The discovery—at just the right time— of an almost complete skeleton of a sheep-sized species of the former, was a fortunate chance that confirmed his inferences based on separated bones (fig. 9). Reflecting on this strange fauna, consisting almost wholly of what he termed pachyderms, he drew an analogy with the present mammalian fauna of Australia, likewise almost wholly of one group, namely the pouched marsupials such as kangaroos. He tantalized his readers by claiming that the analogy was important for "establishing some conjectures" on what the Paris region had been like at the remote "epoch" when the fossil animals had been alive, but he did not specify what the conjectures were.

As in his earlier lecture at the Institut (text 5), Cuvier claimed that it was not difficult to reconstruct the muscles and other features, once the skeleton was reliably assembled; and he inferred for example that these animals had had a short trunk, like a tapir's. It should be emphasized, however, that here and in his other papers on fossils the goal of *reconstruction* remained not only subordinate but also largely unfulfilled: Cuvier's primary aim was simply to "determine" or *identify* the zoological

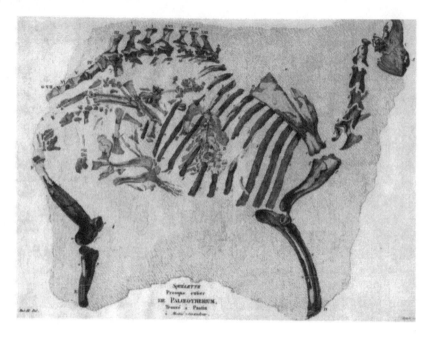

FIGURE 9   A relatively complete skeleton of the *Palaeotherium minus,* found near Paris in 1804: an engraving published by Cuvier the same year.

affinities of fossil bones, and to assign them to their correct place—a new place, if necessary—on the taxonomic map.

In this case, however, Cuvier did in fact write a brief paper to conclude his series on the palaeotherium and anoplotherium, in which he described the procedures he used to reconstruct their skeletons (text 7). The paper was illustrated by engravings of the skeletons of those species for which there was adequate material; the lively poses he gave them leave no doubt about how profoundly he understood the functional anatomy of *any* mammals (see fig. 10). He even suggested in words what the extinct species would have looked like, if fully reconstructed as living animals. But this paper was not published in the *Annales,* and Cuvier's series on the palaeotherium and anoplotherium petered out rather tamely in 1808, without any general synthesis. Only when the papers were reissued four years later in *Ossemens fossiles* did he add his modest verbal and pictorial reconstructions.

Perhaps Cuvier was afraid that if he published his reconstructions in the Muséum's own periodical, his colleagues would accuse him of the speculative bent he criticized so strongly in others. Certainly he *never* published the superb drawings that he also made, probably around this

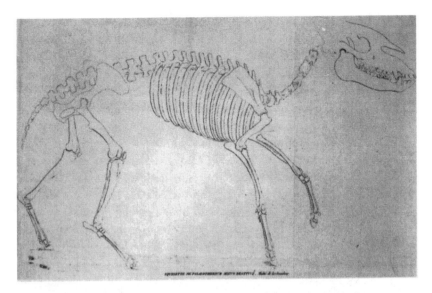

FIGURE 10   Cuvier's published reconstruction of the skeleton of *Palaeotherium minus*, one of the mammals from the gypsum around Paris. Note his careful distinction between the parts for which he had positive evidence, and those he inferred only by analogy (drawn with *dotted lines*).

time, showing both the skeletons and the inferred soft parts of several species, as if the animals had been caught in motion by an X-ray camera (see fig. 11). Whatever Cuvier's failure of nerve may have been—if that was what it was—one of these reconstructions deserves inclusion in this volume, as a clear indication of what he at least aspired to achieve with his work on fossils.[4]

TEXT 7

General summary and reconstruction of the skeletons of the different species [of mammals from the plaster stone around Paris]

HAVING OBTAINED all the pieces of the skeletons of our animals—by the lengthy and laborious analysis that has filled the six preceding papers—and having assigned to each separately its appropriate place, it was then a mat-

4. A revised version of his verbal reconstructions was published in the second edition of his *Ossemens fossiles* (1821–24), with outline sketches of the bodies of four species (both texts and pictures are reproduced in Rudwick, *Scenes from deep time* [1992], pp. 32–37). By this time Cuvier's reputation was unassailable, and he had even more material to substantiate his inferences.

ter of creating a synthesis, bringing them together, making an assemblage out of them, and reproducing before our eyes, if not the whole animals, at least their bony framework.[5]

For this purpose we took the most complete fossil skeletons that we had, as the basis of our work on each genus; and then, searching in the preceding papers for the bones that are missing from the skeletons but that belong to the same species, we have added them on.

We first employed drawings, using only dotted lines—after the example of geographers—for the parts restored by mere conjecture, and continuous lines for those parts copied from the actual specimens [fig. 10]. We then imagined an even more convincing means, which has been successful for certain parts. For example, we possessed a large enough number of separate bones from the feet of *Anoplotherium commune*, that in sorting them by size we were able to reconstruct the four feet, just by substituting what was missing from the bones of one side with replicas in wax of the bones from the opposite side. We have thus brought together all the pieces, after detaching them from the gypsum, by placing them on a bed of clay (because they are too fragile to be assembled otherwise); and we have made them parts of the skeleton as if they had come from the same individual, even though it was necessary to use pieces from perhaps twenty individuals. In this way it has been easier to draw these parts correctly; and the effective reunion of all these bones makes it more striking to the observer that he was obliged to assemble them solely by thought,[6] after having laboriously learned to recognize each of them separately.

This procedure having been very successful for *Anoplotherium commune*, we tried it on the palaeotheriums; and although we were less rich in the bones of that genus, we have also managed to assemble some parts of which we made the same use.

There can thus be no doubt that the drawings that accompany this paper, which offer the general result of our research on the unknown animals whose bones fill our gypsum quarries, represent very closely the skeletons of these animals as they would have been if we had drawn them immediately after their death.

ANOPLOTHERIUM COMMUNE

... Here therefore is the osteology of our animal, completely reconstructed; all the attachments of the muscles are thus given, and the muscles them-

5. [At this point in the revised version of this article, Cuvier added the set of reconstructed body outlines mentioned in note 4.]

6. [That is, by applying anatomical knowledge of the relations between the bones, rather than by direct observation.]

selves can easily be put back in place. Whoever considers the animal thus reproduced will be struck by its heavy build, its short stout limbs, and above all its enormous tail. Except for the size of its limbs, it had much the same stature as the otter, and it is highly likely that it often moved in or on the water, like the otter, especially in marshy places. But this was doubtless not in order to fish there. Like the water rat, the hippopotamus, and all kinds of boar and rhinoceros, our anoplotherium was a herbivore; thus it went in search of the roots and succulent stalks of aquatic plants. Given its habits as a swimmer and diver, it must have had sleek hair like the otter, or perhaps its skin was even semi-bare like the pachyderms of which we have just spoken [in the previous volume]. It is not likely either that it had long ears, which would have impeded it in its aquatic mode of life, and I can readily conceive that in this respect it resembled the hippopotamus and other quadrupeds that live mostly in water.[7]

ANOPLOTHERIUM MEDIUM

...[8] One can see that, whereas the gait of *A. commune* would have been heavy and shuffling when it walked on land, *[A.] medium* must have had agility and grace [fig. 11]. Light like the gazelle or roe deer, it must have run rapidly around the marshes and ponds in which the *A. commune* swam. It must have grazed on the aromatic plants of the dry ground, or browsed on the shoots of the shrubs. Its movement was doubtless not hampered at all by a long tail. Like all agile herbivores, it was probably a timid animal; and large, highly mobile ears, like those of stags, would have warned it of the least danger. Finally, its body was without doubt covered with short hair, and consequently we lack only its color, in order to portray it as it formerly enlivened this countryside [around Paris], where it has been possible, after so many centuries,[9] to unearth its scanty remains. It should be noted in passing that if, thus reclothed in its skin, it had been encountered by some of those naturalists who want to classify everything according to external characters, they would not have failed to rank it with the ruminants; and yet by its internal characters it is quite far from them, and very probably it did not ruminate.[10]

7. [Cuvier's manuscript drawing of his reconstruction of this species (one of the same set as fig. 11) is reproduced in Rudwick, *Scenes from deep time* (1992), fig. 15.]

8. [The final sentence of the preceding summary of Cuvier's specimens of the bones of this species gives a vivid sense of the impact of the *new* specimens he acquired during the course of his work: "The length of this same bone [the femur] could already be conjectured from that of the humerus, and at the moment that I was writing I had just received a complete one."]

9. [Again a very modest expression of their age. See his much bolder guess in text 8, perhaps written a little later than this paper.]

10. [Another criticism of those taxonomists who did not value Cuvier's anatomical expertise as much as he claimed they should.]

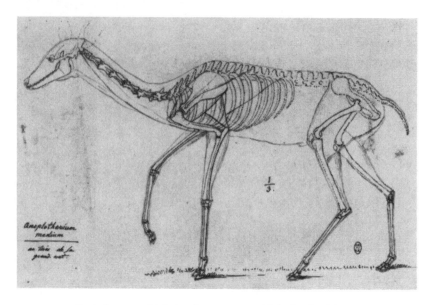

FIGURE 11   Cuvier's pen-and-ink drawing—never published in his lifetime—showing his reconstruction of the "soft" anatomy of *Anoplotherium medium,* one of the fossil mammals from the gypsum around Paris. The bones are drawn in black ink, the inferred soft parts in gray.

ANOPLOTHERIUM MINUS

... If *A. medium* was the roe deer of our region, in the antediluvian world,[11] *A. minus* was the hare: the same size and the same proportions of its limbs must have given it the same degree of strength and speed, and even the same kind of movements.

PALAEOTHERIUM MAGNUM

... There is nothing easier than to represent this animal in its living state, for it is only necessary to imagine a tapir as large as a horse. A naturalist who would have taken the trouble to count its digits would indeed have found one fewer on the forefoot; if he had examined its molars, which so many naturalists today fail to do, he would have found still other differences. But for most people there would only have been that of size; and if one can count on the analogy, its hair was short, or indeed scarcely more than that of the tapir and the elephant.

11. [Cuvier uses the conventional term "antediluvian" to denote the world that existed before whatever "catastrophe" had caused the extinction of the fossil mammals. He was later to claim that the catastrophe was indeed to be equated with the biblical Deluge and equivalent stories in other ancient cultures.]

PALAEOTHERIUM CRASSUM

... This species resembled the tapir still more than the last, for it did not differ even in its size and proportions; and unless its hair was very different I am persuaded that most travelers would have confused these two animals, if they had existed at the same epoch.

PALAEOTHERIUM MEDIUM

... This again was a tapir in appearance, but higher on its legs, and with longer and more delicate feet.

PALAEOTHERIUM MINUS

... If we could have brought this animal back to life as easily as we have re-assembled its bones, we would have thought that what we were seeing running was a tapir as small as a sheep, with light and spindly limbs; that was definitely its form.

PALAEOTHERIUM CURTUM

Finally, there was among these four species a fifth, the *Palaeotherium curtum*, with limbs shorter than in the smallest [i.e. *P. minus*], and almost as stout and squat as in the second [i.e. *P. crassum*]. This would be the extreme in heaviness and ungainliness *[mauvaise grâce]*. But so that this contrast does not surprise us, does not the sluggish wombat crawl, as it were, in the midst of the agile family of jumping kangaroos, climbing opossums, and flying phalangers?

Translated from Cuvier, "Rétablissement des squelettes" (Restoration of skeletons), published in Ossemens fossiles (1812) but probably drafted several years earlier. The passages omitted from the text about each species simply give cross-references to the particular specimens described and illustrated in the papers on the different bones, as the basis for each reconstruction.

7

A POUCHED MARSUPIAL FROM PARIS

Of all the material that was evidently pouring into Cuvier's working space at the Muséum from the gypsum quarries around Paris, one small specimen was so important that it merited a paper of its own (text 8). Marveling at the happy chance of its preservation and discovery, Cuvier referred quite casually to its age as likely to be some "thousands of centuries." Since he was certainly aware by now that the Paris formations were among the youngest known (apart from the loose superficial deposits), the phrase indicates that his implicit sense of the timescale of the *whole* of earth history was quite vast enough to be literally unimaginable.

From the details of the skeleton, and particularly its teeth, Cuvier suspected that this precious fossil had been a marsupial. If correct, the inference was highly significant, because it would show that the tapir was not the only fossil animal from France whose living relatives were found only in the New World and—in the marsupial case—in Australia. So Cuvier staged a risky test of the anatomical principles that underlay all this research. He sacrificed a part of the unique specimen in order to excavate in search of the distinctive "marsupial bones" that support the pouch in living marsupials. In the presence of witnesses who could vouch for his having stated his prediction in advance, he duly found the crucial bones (fig. 12). It was a spectacular vindication of his zoological principles.[1]

---

1. Good modern replicas of the two halves of the little specimen are prominently on display in the Galerie de Paléontologie at the Muséum National d'Histoire Naturelle in Paris. Anyone who sees

With further study Cuvier decided that the specimen belonged to the opossum genus; this meant that the Paris fossil had its closest affinities to animals now confined to America. Research by Geoffroy (who had returned from Egypt in 1801) on the living species of opossums enabled Cuvier to confirm as usual that the fossil species was distinct from any of them. Once again, he drew the moral for "geology": clear "facts" such as this helped to destroy the proliferating "systems" that had been proposed, and thereby cleared the ground for real progress. His spectacularly successful anatomical *prediction* led him to make the bold claim that his own science could aspire to the same kind of quasi-mathematical certainty—at least probabilistically—as more prestigious sciences such as Newton's celestial mechanics and Lavoisier's new chemistry. Tacitly he was seeking to ally himself with his colleagues in the "exact" (physical and mathematical) sciences at the Institut, and distancing himself from Muséum colleagues such as the self-styled "geologist" Faujas.

TEXT 8

———

Memoir on the Almost Complete Skeleton of a Little Quadruped of the Opossum Genus, Found in the Plaster Stone near Paris

IT IS WITHOUT DOUBT a really admirable thing, this rich collection of debris and animal skeletons of an ancient world, assembled by nature in the quarries that surround our city, as if preserved by her for the study and instruction of the present age. Every day some new remains are discovered there; every day adds to our surprise by proving more and more that none of what then peopled the earth in this part of the globe has been preserved on our present soil; and these proofs will doubtless be multiplied to the extent that more interest is shown in them and more attention given to them. In certain beds there is scarcely a block of gypsum that does not conceal some bones: how many millions of these bones have already been destroyed, since the quarries began to be exploited and the gypsum used for building! How many are being destroyed even now by simple negligence, and how many by their small size still escape the eye of even the laborers who are most attentive to collect them! One can judge this by the piece I am going to

---

them there should be impressed by Cuvier's manual dexterity as well as by his understanding of comparative anatomy. (The original specimen, still unique, is kept securely behind the scenes).

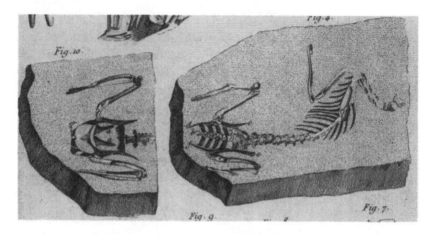

FIGURE 12   Cuvier's drawings of part of a unique specimen of a fossil marsupial from the gypsum near Paris, before *(right)* and after *(left)* he had excavated below part of the backbone *(b)* to expose the marsupial bones *(a, a)* he had predicted.

describe. The lineaments that are found printed there are so faint that one must look at them very closely to grasp them; yet, how precious are these lineaments! They are the imprint of an animal of which we find no other trace, of an animal that, buried perhaps for thousands of centuries *[milliers de siècles]*, reappears for the first time under the eyes of naturalists.

This piece consists of two stones that were collected, and between which, as it were, this skeleton is shared. The first is larger and more complete than the other [fig. 12].[2]

[Here follows a detailed analysis of the skeleton, a lengthy discussion of its affinities, and the conclusion that it belonged to a pouched mammal or marsupial, either of the opossum genus *(sarigue)* of the Americas, or to the dasyure genus of Australia.]

Pouched animals, as is known, are distinguished from all other quadrupeds by two long flat bones, which articulate with the anterior edge of the pubis and serve to support the sides of the pouch, in which these animals carry their young for so long, and which fulfills the extraordinary role of a second womb.

It was necessary to find these bones in this fossil skeleton, or else my demonstration would remain incomplete for those little used to zoological

2. [A single piece of rock had been split open, revealing the skeleton shared between the two surfaces.]

laws and affinities. I noticed that, at the time that the stone was divided into two parts, each bearing an almost complete imprint of the animal, the backbone was split longitudinally; and that its dorsal side remained on the stone on which one sees the head, while the anterior or ventral side was on the opposite stone. I judged immediately that the anterior part of the pelvis must be sunk in the substance of this second stone, under the film that remained at its surface, and which formed part of the sacral vertebrae. So I sacrificed the remains of these vertebrae, between *a* and *b* on fig. [12, right], and between the two sections of the innominate bones, *c d*, *e f*. I excavated carefully with a sharp steel point, and had the satisfaction of exposing to view the whole anterior portion of the pelvis, with the two supernumerary or marsupial bones that I was looking for, in their natural position and wholly similar to their analogues in the opossums.

This operation was done in the presence of some persons to whom I had announced the result in advance, with the intention of proving to them—by the act—the justice of our [i.e. his own] zoological theories, since the true hallmark [cachet] of a theory is unquestionably the power it provides to predict phenomena. I depict this precious piece at natural size and with the most scrupulous exactitude in fig. [12, left]. The marsupial bones are at *a, a*.

From then on, nothing was left to be desired for a complete demonstration of this proposition, already so singular and indeed important, that *there are in the plaster quarries that surround Paris, at a great depth and under various beds filled with marine shells, the remains of animals that can only be of a genus now confined entirely to America, or else of another confined entirely to New Holland.* Up to now the tapir is the only American genus that we have found in fossil form in Europe: the opossum would be the second. As for the genera peculiar to Australasia, they had never been discovered among Europe's fossils.

[Here follows a discussion of the various living animals that have been called opossums.]

To return to my fossil, it was scarcely less curious or less embarrassing for the geologists, whether it was of the New World or of Australasia, that other world still newer to Europeans and above all to naturalists. But the object of my research is to procure light, not embarrassment, for geology. Thus I could only believe I had half fulfilled my task, if I could not manage to destroy this doubt that still remained with me, to determine for myself between these two continents, and to decide finally between the opossum genus and that of dasyures. By reflecting on this problem, by examining and excavating my stone, I had the good fortune to find a means of resolving it.

[Here follows a discussion of the structure of the foot; and the conclusion that the fossil belongs to the genus of opossums.]

Thus the question is decided as far as it can be, and our previous proposition is more rigorously determined and reduces to this: *there are in our [Parisian] quarries the bones of an animal of which the genus is today exclusively found in America.*

[Here follows an enumeration of the eight known living species of opossum, as they are being studied by Cuvier's "learned colleague Geoffroy"; and the conclusion that the fossil belongs to none of them, although it is closest to the mouse opossum *(Didelphis murina).*]

I will not enlarge on the geological consequences of this paper. It is clear to all who are a little acquainted with the systems that relate to the theory of the earth, that it overturns almost everything that concerns animal fossils. Until now one wanted to see in our northern fossils only Asian animals; it was also readily agreed that Asian animals had passed into America and had been buried there at least in the north; but it seemed that the American genera had [not][3] left their native soil, and that they had never extended to the countries that today form the Old World. This is the second proof that I have discovered to the contrary:[4] persuaded as I am of the futility of all these systems, I find myself pleased each time a well-established fact comes and destroys one of them. The greatest service one can render science is to make a clear space there, before constructing anything; to start with all these fantastic edifices that obstruct *[hérissent]* the avenues [of progress], and that hinder from participation all those to whom the exact sciences have given the felicitous habit of acceding to evidence, or at least of ranking propositions according to their degree of probability.[5] With this last precaution, there is no science that cannot become almost geomet-

---

3. [The negative required by the sense is missing.]

4. [The first was the tapir teeth found in France, the living tapir being thought to be confined to South America.]

5. [The striking use of the verb "hérisser" suggests an image of paths overgrown with tangles of thorny shrubs, needing drastic action with shears to clear a way through. The irony intended in the rest of the sentence should be clear.]

rical:[6] the chemists have proved as much in recent times for theirs,[7] and I hope that the time is not far off when one will be able to say as much of the anatomists.

Translated from Cuvier, "Petit quadrupède du genre de sarigues" (Little quadruped of the opossum genus, 1804), omitting the purely anatomical passages.

6. [The allusion is to the geometrical foundations of the celestial mechanics of Newton, which—so it was widely held—had just been perfected by Cuvier's powerful colleague at the Institut, Laplace (*Méchanique céleste*, 1799–1805); it may also refer to Kant's famous denial of the possibility that biological sciences could *ever* share that status (see text 19).]

7. [An allusion to the newly rigorous and quantitative chemistry developed by Lavoisier before his death during the Revolution, continued since then by his collaborators (among them, Laplace).]

8

POPULAR LECTURES ON GEOLOGY

Cuvier's repeated sniping at "geology" and "geologists" must have infuriated his colleague Faujas, the professor of that science at the Muséum, particularly since the criticism was so relentlessly negative. Cuvier might claim, as in his paper on the fossil opossum (text 8), that his goal was to throw light on "geology," not to cause its practitioners embarrassment; but this must have sounded hollow, as long as he so consistently refrained from making any positive suggestions about the geological implications of his research. However, Cuvier's reluctance to be more explicit about his own conjectures was clearly related to what he saw as the disciplinary status of the various sciences. He was concerned above all to promote his own science of comparative anatomy, by showing it was as rigorous as the physical sciences; if it was to be applied—in the matter of fossils—to the speculative area of "geology" or "theory of the earth," the contrast had to be firmly established.

Those disciplinary constraints could be relaxed, however, if he was not primarily addressing his colleagues. In 1805 he gave courses of lectures both at the Athenaeum and at the Collège de France, bearing for the first time the title "Geology"; and his course on physiology at the latter institution was also introduced with lectures on geology.[1] These

---

1. The surviving records of the lectures (see below) make it likely, but not certain, that he spoke about "geology" in three separate courses. The manuscripts need much more detailed study, before

FIGURE 13   A portrait of Cuvier around the time of his first lectures on geology (1805), based on a painting by François André Vincent. The engraver, Miger, also engraved some of Cuvier's plates of fossil bones, and taught the craft to Cuvier himself.

courses, and particularly the first, put him into public view, even more than before, as one of the most prominent men of science in Paris (fig. 13). His audiences consisted mainly of students and the general bourgeois public; indeed, his fellow savants apparently thought his course at the Athenaeum would be a mere popularization of no scientific interest, and learned too late that Cuvier was making novel and important claims.

a definitive account of Cuvier's lectures can be given; what follows is merely a provisional synthesis, which should suffice for the purposes of this volume.

Cuvier's full notes for his courses have not survived. A very brief set of preliminary notes for those at the Athenaeum (text 9) seems to be his first draft of the main claims he planned to make in that course, with a few of the empirical observations he intended to mention in their support. Scrappy though they are, these notes do at least show how he intended *his* "geology"—in contrast to most of his predecessors—to be a science based on *fossils,* as the main evidence for the major changes the earth has undergone. He intended to use them to argue (1) that since the oldest rocks lacked any fossils, life had not existed eternally; (2) that not only had dry land emerged from the sea, but also that the sea had invaded the land; and (3) that there had been a sequence of different forms of life. He then planned to claim—as he had done already in his published papers—that some of these changes had been sudden, and could not have been caused by any physical agency currently at work. The argument would be backed by citing some well-known field observations, most of which he knew only from his reading.

This outline is consistent with a synopsis of Cuvier's subsequent lectures, written by a well-informed Italian member of his audience. Giuseppe Marzari Pencati (1779–1836) was a young nobleman from Vicenza, who was visiting Paris to improve his knowledge of the sciences; in particular he was studying "geology" with Faujas. According to Marzari Pencati, Cuvier followed Buffon in dividing the history of the earth into several "epochs" or periods.[2] There had first been a primal universal ocean, which precipitated the Primary rocks. Life had arisen when continents were also formed. The present continents, Cuvier argued, were not immeasurably ancient, and might have emerged from the sea no more than ten thousand years ago. There was no evidence to support a transformist view of life: human and animal mummies showed there had been no organic change since the time of the ancient Egyptians. There was also no evidence of gradual change between fossil and living animal species; on the contrary, all the evidence pointed to the reality of extinction. In particular, no human fossils were known.

None of this was strikingly original, though Cuvier had not previously adopted so publicly Deluc's well-known views on the recent date of the emergence of the present continents. However, Marzari Pencati also recorded that Cuvier, like Buffon and Deluc before him, expounded the history of the earth as a sequence of *six* periods, and equated that

---

2. The original manuscript is among Marzari Pencati's papers now preserved in the Biblioteca Comunale Bertoliana in Vicenza; the following summary is based on that given in Corsi, *Age of Lamarck* (1988), pp. 182–83.

sequence—at least in broad outline—with the "days" of creation in Genesis. How strongly that theme was stressed is difficult to judge: Marzari Pencati recorded that Cuvier merely noted that "the oldest of books is Genesis, which for whatever reason always tallies with the geological record." This was hardly a novel claim, and certainly not comparable to Deluc's fervent advocacy of the parallel between geology and the creation story: it was simply based on the observation that "the geological record shows us that fossil fish are always found underneath, and mammals above ... never men, the latest and newest creations." [3]

Nonetheless, Marzari Pencati's report of the lectures, as sent to a naturalist in Deluc's native city of Geneva (text 10),[4] shows that Cuvier was perceived by his audience as making an abrupt volte-face: from having been assumed to be on the side of the atheists, he seemed to have switched to the newly powerful party of religious orthodoxy (Cuvier's own views on this point are discussed below).

On the very day that Marzari Pencati wrote that letter about the course at the Athenaeum, Cuvier began his *physiology* course at the Collège de France with—rather surprisingly—five lectures on geology. Another young nobleman and naturalist, Jean-Baptiste Julien d'Omalius d'Halloy (1783–1875) from Liège in the southern Netherlands (now Belgium), wrote a synopsis of the lectures; again, Cuvier's own notes have not survived.[5] Cuvier began with arguments to refute Buffon's claim that there had been a gradual cooling of the globe; he then tried to improve on the older naturalist's attempts to estimate the timescale of whatever changes had in fact taken place; the third part was devoted to a review of Buffon's "theory of the earth," and those proposed before and since his time; and finally Cuvier put forward his own theory.

Cuvier presented himself here as a worthy successor to the greatest naturalist of the Old Regime, the director of the institution from which the Muséum had been reformed; Buffon's theory was the linchpin of

3. As translated in Corsi, *Age of Lamarck* (1988), p. 184. The word "creations" was used generally and casually by naturalists; it was quite compatible with the belief that new kinds of organism had come into being by some unknown natural ("secondary") cause, rather than by the "primary" cause of direct divine action.

4. This brief text is the only one in this volume that is not by Cuvier himself. It deserves inclusion because it gives an important insight into how the religious implications of Cuvier's geology were perceived by the Parisians who attended his lectures; it has frequently been cited and quoted by historians, but as far as I am aware it has not previously been translated into English.

5. The geological lectures were given from 20 Floréal to 2 Prairial, Year XIII (10–22 May 1805). The synopsis by d'Omalius has only recently come to light, and has been summarized by Grandchamp ("Cours de géologie," 1995); it forms a small part of a massive collection of notes on the many lecture courses that d'Omalius followed while in Paris. He became a leading geologist in what later became Belgium.

Cuvier's review of earlier "systems." As for his own theory, it was based on the insufficiency of "slow causes" to account for what he argued had been occasional "sudden changes." But apparently he emphasized that those "revolutions" had *not* been universal or global: on the contrary, they had been local and particular to different "basins." According to d'Omalius, Cuvier noted that they were as yet poorly known, that they would need to be studied in each region separately, and that "it is impossible to assign general laws to them"; the argument probably reflects Cuvier's growing awareness of the complexity of the events recorded by the Parisian formations in which some of his most interesting fossils were being found (texts 7, 8). As for the biblical implications of geology, Cuvier may not have mentioned them at all; at least d'Omalius recorded no such remarks.

Finally, Cuvier seems to have given a separate course on *geology*, during the same year at the Collège de France. Again no notes of his own survive, but a detailed synopsis by yet another auditor—this time an anonymous one—gives a good impression of their content.[6] Evidently Cuvier used much the same material as for his course at the Athenaeum: one was given on the Left Bank of the city, the other on the Right, and there was probably little overlap in their audiences. The following summary therefore incorporates a general evaluation of all his lectures.

Cuvier explained that there were two distinct sources of evidence—organic and inorganic—for learning about the earth's "revolutions." Naturally he gave pride of place to the fossil evidence, which was not only his own field, but also the more novel aspect of the subject. He reviewed the different kinds of fossil in the traditional order of the "scale of beings"—though he had now dropped any overt reference to that ancient concept—from the "highest" animals down to the plants. The mammals, the focus of his own research, therefore came first and were given the most detailed treatment. Cuvier summarized work that he had only just begun to publish in the *Annales du Muséum,* so that there was much here that would have been new even to his colleagues, had they been there to hear it.

By contrast, when Cuvier moved on to consider the inorganic evidence about the earth, his course contained little that was at all original. This was not his science. Indeed he had been scorning "geology" until now as a tissue of unsubstantiated speculations; and his chosen career

6. The manuscript is preserved in the Bibliothèque de l'Institut de France in Paris (MS 2378/6). Its title, "Abrégé des leçons de géologie faites au collège de France, en l'an XIII" (Synopsis of lessons on geology given at the Collège de France in Year XIII), seems to refer to a course distinct from either of those summarized above.

path as a museum naturalist had deprived him of the opportunity to see, at first hand and *in the field,* most of the phenomena that geological theories were designed to explain. But he had picked up what was common knowledge among savants who did deal with such matters. He set his own research into the broader context of a *history* of the earth. It owed most to the traditions of fieldwork, and specifically to the German practice of "Geognosie." As mentioned in chapter 1, geognosy was the careful description of the successive rock formations that composed the earth's visible crust, with relatively little attention to high-level causal theorizing about how the structures described had come into being. However, there was a broad tacit consensus about the kind of earth history suggested by that descriptive work; in the work of theorists such as Deluc, it became explicit (see text 2).

Cuvier adopted this standard model quite uncritically. He outlined a history in which the oldest rocks, the "Primitive" or "Primary" formations, dated from a period before there was any life on earth; they had been precipitated from an initially universal "liquid." Further precipitations had produced further formations, and at a certain point life had begun, as shown by the appearance of fossils, both marine shells and land plants. At least some of these "Secondary" rocks, however, had been formed not by chemical precipitation but by the erosion of older rocks, since conglomerates (solidified gravels) showed rounded pebbles of one rock embedded in another. Still younger formations contained the fossil remains of vertebrate animals; but there were no true human fossils, and man was evidently a very recent arrival on the scene.[7] During all this long history the initially universal "liquid" had been transformed by successive chemical precipitations into the merely salty water of the present seas and oceans; and as its global level fell (with some oscillation), the continents had emerged.

Cuvier's application of the language of "revolutions" to this scenario was also unoriginal, since in itself the term simply denoted major changes in the course of time: as on previous occasions he assumed that the reality of such changes was uncontroversial. It is important to note that he distinguished general revolutions, or changes that had affected the whole globe, from "partial" ones the effects of which had been restricted to particular regions; and he implied that revolutions had tended to become more localized in the course of earth history.

7. The Primary and Secondary formations of geognosy *cannot* be correlated in any simple way with the Precambrian, Paleozoic, Mesozoic, and Cenozoic of modern geology. They correspond better (though not precisely) with the informal distinction, well understood by modern geologists, between "hard-rock" and "soft-rock" regions.

However, Cuvier added to the general idea of "revolutions" the claim that many of the changes had been sudden and violent "catastrophes." But this too was unoriginal, being correctly identified by Marzari Pencati (text 10) as similar to the claims of older savants such as Dolomieu, Saussure, and Deluc. Cuvier referred explicitly to sudden revolutions as having been repeated many times during the history of the earth, but he followed his mentors in focusing on the most recent such event. This was not only the one for which the evidence was clearest, precisely because it was the most recent; it was also of course the one in which the place of man in that history was implicated. As he had suggested in some of his earliest work (text 3), Cuvier clearly regarded the most recent revolution as the one that separated the present *human* world from one that had been essentially *prehuman*. Had there been human beings before that event, he argued, their fossil remains should have been found.

What was novel about Cuvier's presentation on this point, and evidently a surprise to his audiences, was his claim that the last revolution—and hence also the appearance of man—had been very *recent* in geological terms. Since he had used the language of "thousands of centuries" quite casually (text 8) to describe the likely age of the Paris fossils—which in turn he now knew were relatively recent in the history of the earth—it is clear that his conception of the whole timescale was far from being restricted; a claim that only about ten thousand years might have elapsed since the last revolution made that event recent indeed. What mattered about the claim, however, was not so much the figure in itself, but rather its relation to the history of man.

Cuvier had much earlier adopted the metaphor of the geologist as the "antiquarian" (or archeologist) of nature, with fossils as the "monuments" of nature (text 5). More recently, however, he had begun to be involved in the work of those for whom the reconstruction of ancient human history was no metaphor at all, but an everyday reality. He and some colleagues at the Muséum had been co-opted as consultants—on matters of natural history—for the French edition of the celebrated *Asiatick researches*.[8] This was the learned periodical in which British scholars, many of them employed in India, had been publishing their epoch-making work on the translation and interpretation of Sanskrit and other ancient oriental texts. Such studies made Cuvier aware—if he was not already—of the long-running and often acrimonious debates about the chronology of human cultures.

---

8. It was published only a few months after Cuvier's lecture course: Labaume, *Recherches asiatiques* (1805).

The scholarly study of historical "chronology" had been an ideological minefield since the seventeenth century, because it had been used both to support and to undermine Judeo-Christian claims for the uniqueness of the revelatory events recorded in the Jewish scriptures, or Old Testament. Critics of that tradition had argued that its privileged position was fatally undermined by the records of other ancient cultures (particularly the Egyptian), which recorded a far greater human antiquity, of tens or even hundreds of thousands of years; whereas the religious had retorted that the alleged records were clearly no more than myths, legends, or fables.[9] The development of critical methods of textual analysis had tended to favor the latter view, and by Cuvier's day the work of the orientalists had helped consolidate a scholarly consensus. This was, as he correctly summarized it, that there was no good evidence to carry the textual records of *any* human culture any further back than a very few thousand years.

For Cuvier, that helped substantiate what several older naturalists whose work he respected, particularly Deluc and Dolomieu, had already concluded on physical grounds. Like them, he claimed that the continents now inhabited by man had existed in their present form for no more than a few thousand years. Before that time, their surfaces had been on the seabed; if there had been other continents at that earlier period—as the fossil remains of terrestrial plants and animals implied—they had been situated elsewhere, perhaps in parts of the globe now beneath the sea. Cuvier therefore outlined a synthesis between the natural and the textual evidence, and inferred that human cultures dated only from after the most recent revolution, in which the present continents had emerged from beneath sea level.

That conclusion, however, immediately recalled the arguments repeatedly put forward in the prolific publications of the indefatigable Deluc, who made no secret of the fact that *for him* its importance lay in the support it gave to the historicity of the deluge story in Genesis. How much sympathy Cuvier had for that inference is difficult to tell from the evidence now available. In his course on geology at the Collège de France he may not have spoken about it at all: his summary of the various ancient cultures whose records bore on the problem did not even mention the one that was crucial from Deluc's point of view.[10]

9. Knowledge of ancient Egyptian history was still based only on secondhand reports in the classical Greek literature; the hieroglyphic script was not deciphered until the 1820s, partly on the basis of the trilingual Rosetta Stone discovered by the French military expedition to Egypt.

10. It is just possible that it was discussed at the end of the course (the last sheet of the auditor's manuscript is missing).

In any case, for Cuvier—though not in the same way for Deluc—the last or most recent revolution was not unique as a physical event (though it was, necessarily, in relation to the history of man). It had been just one of a series of similar events that had occurred repeatedly, either globally or in localized form, at various times during the earth's history. Revolutions, whether sudden or not, were clearly part of the natural order of things. Cuvier offered few suggestions about their causes, however, evidently because no conjectures seemed satisfactory. His queries about possible explanations for the apparently vast fall in global sea level indicate that he felt in such matters that he—or any other theorist—was merely clutching at straws; but the form of the queries shows unequivocally that he assumed the causes must have been wholly natural and physical in character.

The parallel question of the causes of change in the *organic* world, on the other hand, was highly charged with more problematic implications. Suggestions about the natural transformation (in modern terms, evolution) of species were generally considered—both by their proponents and by their critics—to carry ineradicable implications of philosophical "materialism," and therefore of irreligion. Such speculations had been circulating in Paris for some years, but had been easily dismissed by savants such as Cuvier, on the grounds that they were incompatible with current scientific knowledge. However, in 1800 such "transformist" views had been publicly adopted by one of his own colleagues, in a course at the Muséum. That had made the matter much more serious.

The colleague was Lamarck, already mentioned as the professor of *insectes et vers* ("insects and worms"; in modern terms, of invertebrate animals); he had earlier been a distinguished botanist. But Lamarck regarded himself above all as a wide-ranging "philosopher-naturalist," of a kind that was utterly antithetical to Cuvier's more modern and "disciplinary" conception of scientific work. Lamarck had published his own highly speculative "theory of the earth" three years earlier (*Hydrogéologie,* 1802); and just as Cuvier was giving his lectures Lamarck was publishing an almost equally speculative paper on geology in—of all places—the *Annales du Muséum* ("Considérations sur la théorie du globe," 1805). Yet Lamarck was a colleague far too powerful for Cuvier to attack openly, even in a course for the general public.

In his lectures, therefore, Cuvier seems to have prudently chosen an easier target for his criticism of transformist speculations. De Lamétherie, the editor of the *Journal de physique,* had just published a book expressing evolutionary ideas, which could be ridiculed more easily than

Lamarck's.[11] Cuvier was clearly hostile to the materialistic overtones of current transformist theorizing, but it does not necessarily follow that he regarded species origins as supernatural; certainly he was careful to use neutral language to refer to the causes of the origins of new forms of life, and even of man.

At least in his course at the Athenaeum, Cuvier evidently disappointed the antireligious elements in his audience, by failing to take their side as they had expected. The comments of Marzari Pencati (text 10) make it clear that Cuvier had been regarded as a skeptic in religious matters, like many other Parisian intellectuals, so it was dismaying to that party to find that such a heavyweight savant had apparently abandoned them. Conversely, the side to which Cuvier seemed to have given at least tacit support felt it was politically in the ascendant. The pope had just arrived in Paris to witness Napoleon crown himself as self-styled emperor, thereby consummating the reconciliation between the papacy and the French state, after the anti-Christian campaigns of the Revolution. That event appealed to a nostalgia for the supposed religious certainties of the Old Regime, and reinforced a tendency to equate any kind of modern thinking, not least in the sciences, with the irreligion of the Revolution. So Marzari Pencati thought at first, with some cynicism, that Cuvier's apparent espousal of geological opinions compatible with the creation narrative in Genesis was mere political opportunism.

The records of Cuvier's work, however, seem to confirm what Marzari Pencati concluded on further reflection. Cuvier was not simply trimming his sails to current political winds, but rather taking the opportunity to make his personal views explicit, in a popular forum where he felt he could relax the rigorous standards appropriate to meetings of savants. Nonetheless, the timing of his remarks was probably not coincidental. If the foundations of science—as Cuvier conceived them—were under attack from materialists, so too was the legitimacy of scientific work under attack from religious conservatives. Cuvier may well have felt that he was caught in the middle, between two *equally* unattractive factions. He may have considered it important to use the occasion to defend his kind of

11. Lamarck himself has been taken to be the direct target (Burkhardt, *Spirit of system* [1977], p. 195; Corsi, *Age of Lamarck* [1988], p. 181); but the wording of the report translated here in text 10 suggests it was de Lamétherie's book. In any case, however, Lamarck's ideas were certainly under covert attack. That Cuvier had dropped any explicit reference to the scale of beings is probably due to its use by Lamarck and others, as the basis for their transformist theories: they conceived the evolution of life as having—at least in broad outline—ascended the scale from "lower" organisms to "higher" in the course of time.

science, in the far from liberal climate of Napoleon's regime: he needed to show the general public that the new science of geology—in which his own work on fossils was now deeply implicated—was not necessarily antireligious, either in tone or in its implications. But he must also have felt, equally, that geology was yielding reliable new knowledge that was incompatible with the literalistic biblical interpretations being promoted by some religiously conservative writers at just this time.[12]

TEXT 9

———

Course at the Lycée[13] in Year XIII [1804–5], on Geology

GENERAL PLAN

It is in beds with fossils that there are the strongest proofs that the globe has not always been as it is at present.

STATE OF FOSSILS

Simple fossils[14]—petrified and metallized—embedded in stones etc. They are not sports of nature [jeux de la nature][15]—proved by their texture and chemical composition. They do not form by themselves the beds that contain them.[16]

Prop[osition 1]. The parts that contain no organized bodies at all are the most ancient. Thus organization [i.e. life itself] has not always existed.

Prop[osition 2]. Granitic mountains are not stratified [par couches].[17]

Prop[osition 3]. There have been several successive changes of state, from sea into land, from land into sea, and in one and the same sea.

12. Notably in the famous *Génie de Christianisme* (Genius of Christianity, 1802) by François-Auguste-Réné de Chateaubriand (1768–1848), a prominent Parisian writer, historian, and diplomat—but not a naturalist.

13. [The Athenaeum had been known as the Lycée Republicain until Napoleon turned the French Republic into first a consulate and then an empire.]

14. [That is, those preserved without major change of substance.]

15. [An ancient term (often in the Latin form of *lusus naturae*), denoting the belief that any resemblance between fossils and living animals and plants was purely fortuitous.]

16. [That is, ordinary rocks (apart from limestones, etc.) are *not* all of organic origin, as Lamarck among others had claimed.]

17. [The statement suggests that Cuvier did not accept the common opinion that granite was a rock that had been precipitated from the global "liquid" like any other. The synopsis of his other lectures is ambiguous on this point.]

Prop[osition 4]. Several of the revolutions that have changed the state of the globe have been sudden.[18]

Prop[osition 5]. There have been different ages, producing different kinds of fossils.

Prop[osition 6]. Fossils have often been deposited in tranquil water, and have not been transported at all.[19]

Prop[osition 7]. The causes still active at present, or those taken to be, could not have produced the changes of which traces are found. Neither volcanos, nor the tidal ebb and flow, nor the shifting of the sea toward the west,[20] nor the flooding of rivers etc., nor alluvial deposits [alluvions], nor coral reefs [lithophytes]. Of the retreat of the sea.[21]

MISCELLANEOUS NOTES [22]

Only about 1/1,600 of the diameter of the earth has as yet been penetrated.[23]

The *falun* of Touraine extends over [an area of] nine leagues square [over 700 square miles] with a thickness of twenty feet.[24]

The beds of rolled pebbles, sometimes vertical.[25]

18. [Here as elsewhere, Cuvier's wording makes it clear that a "revolution" is *any* major change, any one of which *may*—or may not—have been sudden.]

19. [This claim was to eliminate the popular explanation that "tropical" animals and plants were found fossilized in northern latitudes because they had been swept there in some kind of violent flood. Cuvier's wording also indicates that he assumed that the past history of the earth had indeed been largely tranquil, not continuously "catastrophic."]

20. [Two processes prominent in Buffon's writings, and those of many other theorists such as Lamarck; the latter process referred to the supposed generalization that the western coastlines of the continents are undergoing erosion while the eastern are shorelines of deposition.]

21. [That is, its final retreat off the present continents, and the consequent emergence of the continents as land areas, which constituted the most recent "revolution."]

22. [All the notes below refer to observations that were well known to naturalists in Cuvier's time (he may even have seen one or two for himself on his travels). They give at least a sense of the way he planned to support his "propositions" with empirical evidence.]

23. [That is, even the deepest mines only scratch the surface of the globe. Cuvier probably planned to argue that causal theories should therefore not be based on speculations about the deep interior, about which almost nothing is known.]

24. [A deposit of sands with fossil shells, covering a vast area of the region around Tours. Cuvier probably planned to use it to impress his audience with the magnitude of the apparently recent changes in geography.]

25. [The vast spreads of coarse gravel in some of the Alpine valleys (which Cuvier may have seen on his travels to Italy and back) seemed to bear no relation to the rivers that now flow there. Some much older deposits of the same kind (in modern terms, conglomerates) had been found in the Alps tilted into a vertical position, thus providing some of the strongest evidence for the reality of major crustal movements.]

Shells are found at 2,400 feet [above sea level] on Etna—at 14,022 feet on the Andes—on the Pyrenees.[26]

Table Mountain is granitic [and] most of the mountains to the north likewise.[27]

The trees buried in the peats of Holland and Westphalia lie from southeast to northwest.[28]

Translated from a small manuscript in Cuvier's hand, MS 3111, Bibliothèque de l'Institut de France, Paris. It is reproduced here in its entirety. The original French text is transcribed in the appendix. Cuvier's "propositions" are numbered for convenience of reference.

TEXT 10

—

[Cuvier's lectures at the Athenaeum]

Paris, 20 Floréal, Year XIII [10 May 1805]

Cuvier has given a course at the Athenaeum which he terms *Geology*, but which I'd rather call *Comparative anatomy applied to geology*. No one, I believe, knows the osteology of all the classes of vertebrate animals better than that great man; no one could make better use of it in the identification of fossil mammals, reptiles, birds, and fish; and from that identification one could scarcely draw consequences more legitimate, more important, or more general. This is perhaps the only part of geology on which one can write without copying the Genevans, or at least without helping oneself to their materials.[29]

It wasn't without astonishment—knowing he is hardly very devout *[très dévot]*—that I heard him wanting to have all his corollaries[30] in the service

26. [In the absence of any strong concept of crustal movement, this was usually interpreted as evidence for the vast scale of the gradual fall in sea level during the history of the earth.]

27. [The meaning of this note is obscure.]

28. [The reported orientation of tree trunks buried in peat (some of which Cuvier may have seen on his travels) was commonly regarded as evidence that a tsunami-like "deluge" had swept over Europe from the southeast.]

29. [An allusion to the outstanding fieldwork of Saussure and those he had inspired to continue his study of the Alps and other mountain regions. Possibly Deluc—also a Genevan until he moved to England—was also in the writer's mind; anyway the comment was flattering to his Genevan correspondent, and to the societies with which he was associated.]

30. [That is, the inferences Cuvier drew from his fossil research.]

of proving the opinion of Dolomieu, Saussure, and the two Delucs, about the newness of our present continents. The Holy man *[le Saint homme]* doesn't assign them even ten thousand years. The new work on *organized beings [êtres organisés]*[31] was ridiculed as it deserved; its formation of different organs by habit was joked about. ... And whatever consideration *[ménagement]* the pleasant and learned materialist who is its author may deserve, a CUVIER COULD TAKE THE LIBERTY OF JOKING, WHERE ANIMALS WERE CONCERNED.

It seemed to be in good faith, and it must be admitted that it's an appalling loss for the atheists, who were proud in advance of what the great Cuvier would deduce from his zoology and his zoological geology. Calculating from his former prejudice, they had already announced, as his, some opinions that were scarcely his at all. One can truly say that they have quite an adversary there ... and if he is such, he's one who is indeed formidable, and well worth a Gall.[32]

To decide *whether he is or is not in good faith,* I will add for you that, having started his course when the pope had just arrived [in Paris], and having set out for us in the beginning his altogether Mosaic *[Moysienne]* opinion,[33] I thought at first that the good devout man was simply on the lookout for a cardinal's hat. This conjecture was not extravagant in a city where all are flatterers, and at a time when everyone is changing position. If one zoologist at the Muséum could be elected grand chancellor of the Legion [of Honor], why shouldn't another be made a cardinal?[34]

But I promise you I've gone back on that suspicion completely, and I believe he's sincere, apart from the *jokes,* for he led us step by step without any sophisms, as far as I could tell. It's true he told us things that are clearly

31. [Probably, by the wording, a reference to de Lamétherie's recently published *Considérations sur les êtres organisés* (Considerations on organized beings, 1804), rather than to Lamarck's earlier *Recherches sur les corps vivans* (Researches on living bodies, 1802).]

32. [Cuvier was deeply opposed to the controversial physician Franz Joseph Gall (1758–1828), not only for his materialistic ideas on the relation between mind and brain, but even more for his promotion of what Cuvier regarded as the pseudo-science of phrenology. His later report to the Institut on Gall's work (1808) attacked it on methodological grounds not unlike his criticisms of the speculative "geologists."]

33. [That is, opinions compatible with a literal reading of the early chapters of Genesis, which were traditionally taken to have been written by Moses himself. As mentioned in the introduction to this chapter, it is not clear from the surviving records that Cuvier did more than note the similarity between the order of events in geological history and those in the creation narrative, without suggesting any reason for the parallel.]

34. [A suggestion not meant to be taken seriously, given the fact that Cuvier was not even a Roman Catholic layman, let alone a priest! The allusion is to Cuvier's Muséum colleague Lacépède, who had accepted this position from Napoleon in 1803.]

wrong, when he wanted to go farther down than the *thin crust with fossils.*[35]
He knew nothing down there; he had made a world in his museum *[cabi-
net]*;[36] he built [mountain] ranges and let us hear how those that are made
of Primitive limestone had been emplaced; as an example he managed to
cite for us Mont Perdu in the Pyrenees, which as every geologist *[géologue]*
knows is completely shelly.[37] But all that doesn't affect his conclusion about
the present continents, which is quite independent; for that's only about
the monuments existing in *formations,* in the knowledge of which he's a
great man: it's solely from those that he's deduced it.

I took notes all through the course; I'm keeping them very carefully, because facts of the greatest importance are recorded there, which have not
yet been published. I believe it would please you if I sent a copy for you and
for the society. For that purpose I've given them to a copyist, for being a
matter of fifty-two pages in folio it's impossible for me to copy them myself.
He'll deliver them before I leave Paris and I'll have the honor of sending
them to you. It was only the general public *[gens du monde]* that followed
this course at the Athenaeum: the geologists would have deliberately subscribed to it if they'd known it would be as interesting as he made it, but
they laughed about it at first. Now I'm asked on all sides for my notes, but I
haven't got time to satisfy everyone.

*From a letter from Marzari Pencati to Henri-Albert Gosse (1753–1816); translated in full
from the excerpts published in Plan, Henri-Albert Gosse (1909), pp. lxxxii–lxxxiv. As in
texts 1 and 2, an informal style is adopted in this translation, to remind the reader that the letter was not intended for publication. Gosse was a leading member of the Société de Genève
pour la Physique et l'Histoire Naturelle (founded 1791), and a founder (1803) of the smaller
but more active Société de Naturalistes de Genève.*

35. [That is, that Cuvier was not familiar at first hand with the vast thicknesses of lower (and
therefore older) rocks, with few or no fossils, but only with the relatively thin "crust" of uppermost
(and therefore youngest) deposits from which his vertebrate fossils came.]

36. [That is, his high-level theorizing was based only on his study of specimens in museum collections, rather than on fieldwork.]

37. [Louis François Elisabeth Ramond, baron de Carbonnières (1753–1827), had reported that
the summit of Mont Perdu, one of the highest peaks in the Pyrenees, consisted of a limestone full of
fossil shells. As such, it was a celebrated and problematic exception to the general rule that the central ridges of mountain chains were composed of granites or other "Primitive" rocks. (The published
French text of this letter reads "Mont Perdu ou Piranèse," which is probably a mistake in transcrip-
tion for "Mont Perdu aux Pyrénées.")]

9

A REVIEW OF FOSSIL PACHYDERMS

C uvier's excursion into the realm of popular exposition was consolidated the year after his first lectures on geology: the galleries of the
Muséum were reopened to the public for the first time since the height
of the Revolution, and his own *new* gallery of comparative anatomy gave
concrete and visible form to his conception of his science. At the same
time the publication of his research in the *Annales du Muséum,* directed at
his fellow naturalists in France and beyond, continued unabated. There
were, for example, papers on the fossil bones of bears, hyenas, and rhinoceros, all distinct from their respective living species. More striking still
was Cuvier's analysis of the celebrated "Ohio animal," known for more
than half a century and the focus of continual debate among naturalists
throughout the scientific world. Cuvier gave an authoritative analysis of
its anatomy; claimed it was a distinct genus, much further from living
elephants than the mammoth (though it had been repeatedly confused
with that species); named it *Mastodon;* and distinguished several species.

In the papers he published in 1806, Cuvier began to be more explicit
about the possible character of the event that, he believed, had wiped out
all these fossil species. There was no grand speculation, just cautious inferences based on concrete evidence. In his rhinoceros paper Cuvier discussed the significance of the well-known report by Pallas, dating from his
exploration of Siberia many years earlier, that he had found a fossil rhinoceros with some of its skin preserved in ice or frozen ground. Cuvier

regarded this as conclusive evidence that the animal had not been swept in from elsewhere, but had lived where it was found. But he argued further—as he had in his lectures—that it must have been overwhelmed *suddenly* for any of its soft parts to have been preserved: its species must have perished not by "slow and insensible changes, but by a sudden revolution."

Cuvier also published a new and greatly enlarged version of his earlier paper (text 3) on living and fossil elephants, at the conclusion of which he reviewed its geological implications (text 11). This was followed by a similar but broader review of his papers on *all* the fossil pachyderms, including those on the rhinoceros, hippopotamus, tapir, and mastodon (text 12). These summaries were clearly intended to present the evidence on which the character of the last revolution could at least be circumscribed.

A new feature of Cuvier's argument referred to specimens he may have acquired only recently. He emphasized that the fossil bones were often found mixed with *marine* debris, and that some even had oyster shells and other marine organisms cemented to them, so that they had evidently been lying on the seafloor for some time. However, their frequently perfect preservation showed that the animals had lived where the bones are found, and that the carcasses had not been swept in from elsewhere, and certainly not from distant parts of the earth. Since the bones were confined to loose superficial deposits, and were not covered by thick or extensive strata of marine origin, Cuvier inferred that the marine incursion must have been relatively brief in geological terms. Since they were found only at low altitudes, the inundation had probably not reached higher ground. Above all, Cuvier claimed that the climates of the regions where the bones were found had been similar *before* the "catastrophe" to what they had been since.

Of course this picture of a transient marine incursion onto the continents had some formal similarity to the traditional image of the biblical Flood or Deluge. However, it is unlikely that Cuvier was trying to lend covert support to the historicity of Genesis, except in the sense that his conclusion would give equal support to other ancient accounts of the same kind. It is clear that he believed that *all* such records were highly garbled legends, with only a small core of historical veracity. So, for example, Cuvier's inundation had *not* covered the whole of the earth's surface—as a literal reading of the Noah story would assert—but only the low-lying regions. It should also be noted that Cuvier's new conception distanced him from Deluc, who regarded the last revolution as having in effect caused a permanent exchange in the positions of continents and oceans, rather than being merely a *transient* marine incursion.

On the other hand, these texts make it clear that Cuvier was trying again to close the door against any transformist interpretation of his evidence. He argued for example that the Indian elephant could not be the descendant of ancient elephants that escaped the catastrophe by being on higher ground. Perhaps he knew his argument was weak at this point, for he bolstered it with a quite different one, namely the agreed lack of any difference between the animals found mummified in the tombs of ancient Egypt and their living counterparts. As mentioned earlier, Cuvier had been one of the authors of the official report on the mummified animals brought back from the Egyptian expedition, which had concluded that they all belonged to living species (Lamarck had been another of the authors, and must have agreed with that conclusion). More recently Cuvier had published a paper in the *Annales du Muséum* on the special case of a mummified bird (the sacred ibis) that had appeared to be an exception; he showed it was not, simply because its living counterpart had been misidentified (see figs. 23, 24). Its wider significance was left implicit in that paper, but it is clear that Cuvier regarded the case as eliminating any possible use of the mummified animals as evidence for the transformation of species in the course of time.

TEXT 11

——

General Results of This [Natural] History of Fossil Elephants

THE DETAILS with which we began have thus shown us that the fossil elephant bones have much resemblance to those of the elephant living today in India. However, we have just seen that almost all of the bones that it has been possible to examine, and to compare precisely with those of the living elephant, presented noticeable differences, greater than for example those between the bones of the horse and the ass. We have concluded from this that these two elephants are not at all of the same species.

This conclusion might not appear completely demonstrated, if it referred only to this single fossil animal, considering that the differences noted are not of very great importance. But it gains force when one sees that the species whose bones usually accompany the elephant's, such as rhinoceros and tapirs, differ still more from their living relatives; and that some, such as the various mastodons, have no known living relative.

The first article[1] has shown us that the fossil elephant bones are usually found in the unconsolidated *[meubles]* and superficial layers of the earth, and most often in alluvial deposits *[terrains d'alluvion]* that fill the floors of valleys or border the beds of rivers. They are almost never there alone, but are mixed with the bones of other quadrupeds of known genera, such as rhinoceros, cattle, antelopes, and horses; and often with the remains of marine animals, such as shellfish, or others of which parts are even attached to them. The positive testimony of Pallas, Fortis, and many others allows no doubt that this last circumstance is often the case, although not always so. We ourselves have at this moment before us a fragment of jawbone covered with millepores and small oysters.

The beds that overlie the bones of elephants are not of great thickness; they are almost never of a stony nature. The bones are rarely petrified, and only one or two examples can be cited in which they were embedded in some shelly or other stone. Often they are simply accompanied by our common freshwater shells; in this respect, as well as in regard to the nature of the soil, the resemblance between the three localities for which we have the most detailed descriptions—namely, Tonna, Canstatt, and the Forest of Bondy—is indeed very remarkable.[2] Everything thus seems to indicate that the event *[cause]* that buried them was one of the most recent that has contributed to changing the surface of the globe. This was nonetheless a physical and general event. The bones of fossil elephants are too numerous, and found in too many deserted or even uninhabitable regions, for us to be able to suspect that they were taken there by human beings.

The beds that contain the bones and overlie them show that this event was aqueous, or that it was water that covered them up; and that in many localities this water was almost the same as that of the sea today, since it sustained roughly similar organisms. But it was not this water that transported them to where they are. The details of this same first article show that these bones occur in almost all the regions that naturalists have traversed. An irruption of the sea that would have carried them only from the areas where the Indian elephant now lives could not have spread them so far, nor dispersed them so uniformly. Besides, the inundation that buried them did not rise over the major mountain chains, for the beds that it deposited and that contain the bones are found only in low-lying plains. Thus one cannot see how the carcasses of elephants could have been transported

---

1. [The first part of the monograph (not translated here), on the geographical distribution of finds of fossil elephants.]

2. [Tonna is near Gotha; Canstatt was just outside Stuttgart (and is now a suburb); Bondy was a few miles outside Paris (and is now likewise a suburb).]

to the north [i.e. to Siberia], over the mountains of Tibet and the chains of the Altai and the Urals.

Moreover, these bones are not at all rounded: their ridges and apophyses are preserved, and have not been worn down at all by friction; very often the epiphyses of those that had not yet completed their growth are still in place, although the slightest effort is enough to detach them. The only alterations that can be noticed come from the decomposition that they have suffered from their burial in the earth. Nor can it be claimed that whole carcasses were transported violently. It is true that in that case the bones would have stayed intact; but they would also have remained articulated and would not be scattered. Besides, the shells, millepores, and other marine products that are attached to some of them prove that they remained at least for some time exposed and detached, at the bottom of the liquid that covered them. Thus the elephant bones were already in the places where they are found, when the liquid came and covered them. They were scattered there, just as the bones of horses and other animals that live in our country—the carcasses of which are widespread in the fields—can be scattered.

Everything therefore makes it extremely probable that the elephants that furnished the fossil bones inhabited and were alive in the countries where their bones are found today. Thus they can have disappeared from there only by a revolution that made all the existing individuals perish, or by a change of climate that prevented them from reproducing there. But whatever this event may have been, it must have been sudden. The bones and ivory that are so perfectly preserved on the Siberian plains are so only because of the cold that freezes them there, or that in general arrests the action of the elements on them. If this cold had only come slowly and by degrees, these bones—and with even stronger reason the soft parts in which they are sometimes, even if rarely, still covered—would have had time to decompose like those found in hot and temperate countries. In this way all the hypotheses of a gradual cooling of the earth, or of a slow variation in either the inclination or the position of the axis of the globe, collapse by themselves.

If the present Indian elephants were the descendants of ancient elephants that had taken refuge in their present climate, ever since the catastrophe that destroyed them in others, it would be impossible to explain why their species has been destroyed in America, where debris is still found to prove that they once existed there. The vast empire of Mexico offered them enough high ground to escape an inundation as little elevated as it must be assumed to have been, and the climate there is warmer than is necessary for their temperament. Moreover, we have shown that the mountains of the isthmus of Panama would not have been an obstacle to their passage into South America.

The various mastodons, the giant tapir, and the fossil rhinoceros lived in the same countries, in the same areas as the fossil elephants, since their bones are found in the same beds and in the same condition. One cannot imagine a cause that would have made some perish while sparing the others. Nevertheless, it is quite certain that the animals mentioned first no longer exist, and in regard to them there can be no argument, as we shall show in the chapters on them. Thus everything concurs to make us think that the fossil elephant is, like them, an extinct *[éteinte]* species, although it resembles a species existing today more than they do.

Translated from Cuvier, "Éléphans vivans et fossiles" (Living and fossil elephants, 1806), pp. 265–69.

TEXT 12

—

General Summary of the [Natural] History of Fossil Bones of Pachyderms from the Superficial Deposits and Alluvium

THE SUPERFICIAL DEPOSITS *[terrains meubles]* that fill the floors of valleys and cover the surface of large plains have thus furnished us, just in the orders of pachyderms and elephants, with the bones of eleven species: namely, one rhinoceros, two hippopotamuses, two tapirs, one elephant, and five mastodons.

All these eleven species are today absolutely foreign to the climates in which their bones are found. The five mastodons alone can be considered as forming a separate and unknown genus, although very close to that of the elephant. All the others belong to genera still existing today in the tropics. Three of these genera—the rhinoceros, hippopotamuses, and elephants—are only found in the Old World; the fourth—the tapirs—exists only in the New World. The same distribution does not apply to fossil bones: the bones of tapirs have been unearthed in the Old World, and some elephant bones have been found in the New World. These species, although belonging to known genera, nonetheless differ noticeably from known species; they must be considered as distinct species, and not just as varieties.

This point cannot be the subject of any argument in the case of the small hippopotamus and the giant tapir. It is even more certain for the fossil rhi-

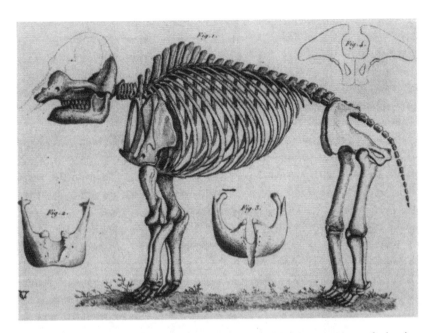

FIGURE 14 Cuvier's reconstruction of the skeleton of the largest species of what he named the "mastodon," previously known as the "Ohio animal" and often erroneously equated with the mammoth.

noceros. Although a little less evident for the elephant and small tapir fossils, there are nevertheless more than adequate reasons to convince the trained anatomist. Finally, the large hippopotamus is the only one of these eleven fossil quadrupeds for which there are not enough specimens to be able to say positively whether or not it differed from the hippopotamus living today.

Of these eleven species, only one, the large mastodon, had been recognized before me as an animal lost [perdu] to us [fig. 14]. Two others, the rhinoceros and the elephant, had indeed been identified as to genus, but I am the first to have shown their specific differences with some exactness. Seven—namely the small hippopotamus, the two tapirs, and the four lesser mastodons—were entirely unknown before my researches. Lastly, the eleventh, the large hippopotamus, at present remains subject to some doubt.

Such is the osteological conclusion of this first part of our work.[3] Such are the different degrees of certainty to which we have been able to bring the various propositions of which this result is composed.

3. [At this stage Cuvier intended to reissue his papers on the fossil pachyderms as the first volume of a three-volume work.]

As for the *geological* result, it consists mainly of the following remarks. These different bones are buried almost everywhere in roughly similar beds; they are often mixed there with some other animals likewise fairly similar to those of today. These beds are generally unconsolidated, whether sandy or silty, and always more or less close to the surface. It is therefore probable that these bones were buried in the last or one of the last catastrophes of the globe.

In a large number of places, they are associated with the accumulated remains *[dépouilles]* of marine animals. But in some less numerous places there are no such remains: sometimes the sand or silt that covers them even contains freshwater shells. No well-authenticated case shows that they are ever covered with regular stony strata filled with marine shells, or— consequently—that the sea remained long and quietly over them.

The catastrophe that buried them was thus a major but transient marine inundation. This inundation did not cover the high mountains, for no deposits analogous to those that contain the bones, nor the bones [themselves], are ever found there—not even in the high valleys, unless in some of the warm parts of America. The bones are neither rounded nor assembled in skeletons, but scattered and partly broken. Thus they have not been carried far by the inundation, but were found by it in the places where it buried them; just as they would have been there, if the animals from which they come had lived in these places and had died there successively. Thus, before this catastrophe, these animals lived in the [same] climates where their bones are now unearthed. It is this catastrophe that destroyed them there; and since they are no longer found elsewhere, it must indeed have annihilated *[anéanti]* the species.

Thus the northern parts of the globe once nourished species belonging to the genera of the elephant, the hippopotamus, the rhinoceros, and the tapir, as well as that of the mastodon; genera of which the first four now only have species that live in the tropics, and of which the last has species nowhere. Nonetheless, nothing authorizes us to believe that the tropical species are descended from these ancient northern animals, transported either gradually or suddenly toward the equator. They are not the same; and we shall see, from an examination of the oldest mummies, that no established fact authorizes us to believe in changes as great as those that it would be necessary to assume for such a transformation, especially in wild animals.[4]

4. [An allusion to the argument, usual among naturalists who rejected the notion of transformism, that the range of variation of a species under domestication (e.g. the range of different breeds of dog) was far greater than any species would ever show under natural conditions in the wild.]

Neither is there is rigorous proof that the temperature of northern climates has changed since that epoch. The fossil species differ no less from living species, than certain northern animals differ from their southern relatives: for example the Siberian [arctic] fox *(Canis lagopus)* from the Indian and African jackal *(Canis aureus).* They could thus have belonged to much colder climates.

These results, already largely pointed out in the article on the elephant [text 11], all seem to me to be rigorously deduced from the facts set out in this first part.

They will be confirmed by the bones of carnivores, ruminants, and others, found in the same superficial beds; but before speaking of those, we are going to deal with the *pachyderms* embedded in regular stony beds, and covered by regular marine strata.[5] These belong to a much more ancient epoch than those we have dealt with hitherto; and we shall also see that they differ far more than them from all animals alive today. They are indeed those that will seem to reappear in this work as a wholly new creation.[6]

Translated from the concluding section of Cuvier, "Dents du genre des mastodontes" (Teeth of the mastodon genus, 1806).

5. [This final paragraph was not published until the text was reissued in *Ossemens fossiles* (1812). It acted there as a link between the volume on the pachyderms from the superficial deposits, and the following (third) volume on the mammals from the much older formations around Paris; the carnivores etc. from the superficial deposits were relegated to the fourth and final volume. (The paragraph appears in Cuvier's manuscript of this text and was evidently written at the same time: MS 631, Bibliothèque Centrale, Muséum National d'Histoire Naturelle, Paris.)]

6. [It is noteworthy that here, where Cuvier does use the word "creation," it refers not to divine activity but to his *own* work as, in a sense, bringing the extinct animals *back* to life.]

A REPORT ON ANDRÉ'S THEORY

OF THE EARTH

———————————

Cuvier had now put forward *two* distinct explanations for the sudden extinction of the large fossil mammals, but the discrepancy between them does not seem to have perturbed him. He did not show how the transient marine incursion he invoked to explain most of the fossil occurrences might have been related to the sudden refrigeration he had inferred to account for the rhinoceros preserved in Siberian ice. However, the latter explanation was reinforced the following year (1807), when reports reached Paris of a similar discovery of a frozen mammoth. As with the megatherium more than a decade earlier (text 4), Cuvier had no specimens to study, only a report; but in this case even a brief verbal description contained important information. He interpreted the description of the thick hairy coat of the mammoth as decisive evidence that it had been adapted to the same arctic climate in which its remains had been found. In that respect it confirmed his general conclusion that in each region the climate before the last "revolution" had been the same as it was now. The event itself—whether a sudden drop in temperature or a sudden incursion of the sea—had been drastic in terms of causing widespread extinctions, but it had been a *transient* event; afterward the physical conditions at the earth's surface had reverted more or less to what they had been before.

Apart from remarks such as these, scattered through his various papers, and of course his summary of fossil pachyderms (text 12), Cuvier had still not discussed the strictly geological implications of his work in terms appropriate to an audience of his fellow savants. At just this time he seized an opportunity to do so. One of the works that was submitted to the Institut in 1806, in the hope of receiving its formal endorsement before publication, was a book by the elderly naturalist Noël André (1728–1808). This put forward yet another "system" for the "theory of the earth." Naturally enough, Cuvier made sure he was appointed to the committee that was to evaluate the book; indeed as "rapporteur" it was he who drafted the report, and the two other members probably played only a minor role in its composition.[1] It was published in the *Journal des mines* and, with minor differences and probably slightly later, in the Institut's own *Mémoires;* in any case it would have been widely read. An English translation—of rather poor quality—was published in the *Philosophical magazine* two years later.

Cuvier's comments on André's book itself were confined to the last part of the report, for he used it quite explicitly as a peg on which to hang his own general reflections on the practice of geology (text 13). Properly conceived, he asserted, geology could be as precise as other physical sciences. It should be concerned particularly with the accurate description of "facts" or observable geological phenomena. By contrast, Cuvier poured scorn on the whole genre of writing to which the book under review aspired to contribute. The proliferation of "systems" purporting to explain the "theory of the earth" had become, in his view, a case of too many theories chasing too few facts; what was needed was a lot more "positive facts," rather than yet more theoretical "systems." This was a theme familiar and congenial to his audience at the Institut; and Cuvier allied his project with those of his more powerful colleagues there, by identifying these geological "systems" with a recrudescence of the Cartesian spirit, which the Newtonian ought to have banished forever.

Cuvier brought Lamarck into the picture, deftly praising him as a scrupulously thorough taxonomist while implicitly condemning his speculative building of a transformist "system." His senior colleague's work on the fossil shells from the strata around Paris had shown that very few were of living species, which was a conclusion that supported Cuvier's own inferences about the fossil mammals. Cuvier urged the Institut to

1. The manuscript of the report (MS 3160, library of the Institut de France, Paris) is in Cuvier's hand; some quite extensive additions to the initial text may represent his responses to comments by the other members of the committee, or by others who heard him deliver the report at the Institut.

encourage detailed factual research of that kind, and to discourage the premature erection of grand theories. Above all, he sketched a research *program* for geology, focusing particularly on the role of fossils in the strata, and on the relatively young (but less thoroughly examined) formations in which fossils were most abundant. Only when such an agenda had been vigorously pursued, he concluded, should "the great problem of causes" be tackled.

In line with his prescription for real progress in geology, Cuvier praised the factual component of André's work—the author had done extensive and arduous fieldwork in several mountainous or hilly regions in western Europe—while maintaining that the Institut should decline to pronounce on the theoretical component. However, he implied that his own ideas about the character of the most recent revolution were in fact quite similar to André's (and also to Deluc's and Dolomieu's): that is, that the surfaces of the continents owed their present form to a relatively recent and sudden event. But in Cuvier's view such private opinions had to be sharply separated from what the premier scientific body should publicly endorse as reliably established knowledge.

The significance of the research agenda that Cuvier proposed for geology can hardly be exaggerated. It represented a major volte-face on his part, yet a legitimate one. Earlier he had dismissed geological "systems" as so many castles in the air, and proposed in effect a total moratorium on such theorizing until there was more evidence to go on. Now, without effacing that sharp and standard distinction between all-explanatory "systems" and reliably established phenomenal "facts," he tacitly acknowledged the possibility and value of constructing lower-level generalizations, based on careful fieldwork. Earlier he had characterized the prehuman history of the globe as one of chaotic disorder and confusing "revolutions." Now his suggested agenda for geological research showed that he acknowledged that all was not as disorderly as it had seemed. A thorough study of specific classes of phenomena could bring at least some local order out of apparent chaos. In particular, he emphasized the importance of attending to the Secondary formations, rather than the Primary ones on which the science of geognosy had focused; and specifically of learning more about the sequences of Secondary rock formations *in relation to their fossils.*

TEXT 13

—

Report of the National Institute
(Class of Physical and Mathematical Sciences) on the Work by Mr.
André Entitled *Theory of the present surface of the earth*

THE CLASS HAS CHARGED Mr. Haüy, Mr. Lelièvre, and me to report to it on a manuscript work by Mr. André, formerly known under the name of Fr. Chrysologue de Gy, entitled *Theory of the present surface of the earth* [*Théorie de la surface actuelle de la terre*, 1806].[2]

Since this is the first notable occasion that has yet presented itself, to speak to the Class on geological matters, it is perhaps not inappropriate to set out first some general reflections on the way in which a group [*compagnie*] such as ours can and should envisage this kind of research.

The natural history of inorganic bodies, commonly termed brute matter or *minerals*, divides into two principal branches. In the first, each of these bodies is examined in itself and in its chemical and physical properties; it is assigned its distinctive characters and its place in the general method.[3] This part has more specifically retained the name of *mineralogy*. Cultivated almost always by good people, it has now reached a degree of precision and exactitude at least equaling that of all the other physical sciences.[4]

The other branch of the [natural] history of minerals has as its object the reciprocal position of their different species and of the [rock] masses composed of one or several of them. It is this branch that teaches us which materials form large tracts of country, and which others are restricted to gaps or fissures in the former. It tells us which substances form respectively the major ranges, the lesser mountains, the hills, and the plains. Above all, it is concerned with the superposition of minerals, and teaches us to distinguish those that always support the others from those that always surmount them — or, in a word, the order followed by their different beds [*couches*]. It is given the names of *geology, geognosy,* or *physical geography,* according as the research is pursued more or less profoundly.[5]

2. [René-Just Haüy (1743–1822) was a priest, a distinguished crystallographer and mineralogist, and Daubenton's successor as professor of mineralogy at the Muséum. Chrysologue de Gy had been André's name before the Revolution, when he was a Capuchin monk.]

3. [That is, in a general classification. The same term was used in botany and zoology; classification was the same practice in all three branches of natural history.]

4. [An allusion in particular to Haüy's work in bringing great quantitative precision to the study of the crystal forms of minerals.]

5. [Cuvier's point becomes clearer if the scale of profundity is reversed: a study of given phenomena constitutes "physical geography" if it is restricted to the description of surface outcrops

It is clear that this is a science capable of as much exactitude as mineralogy proper. For it to acquire that quality, it need only be treated as all the natural sciences ought to be; that is to say, by establishing individual facts with care, and deducing general conclusions only when these facts have been collected in sufficient number, always observing the rules of a rigorous logic. It is also clear that this science comprises a part of the natural history and knowledge of the globe no less indispensable than ordinary mineralogy. It is to the latter, as the [natural] history of climate, of soil, and of the proper description of each plant, are to botany.[6]

Its utility to society—if it is ever to be well done—would be no less evident. By it one would be guided in the search for minerals; it would give one the means to foresee the details and the expense of an infinity of [public] works, which today can only be known by experience [i.e. trial and error]. Thus our engineers were recently unable to calculate the cost of an underground conduit to replace the Marly machine, because they did not know the nature of the formation [terrain].[7] Geology would have told them that at that location only chalk would be encountered.

Miners have more interest than other artisans in possessing this kind of knowledge, and have made a particular study of it in relation to the class of minerals that they pursue. They have identified the character of mountains with metallic veins, and they recognize perfectly the regions where they have nothing to search for and those that could be favorable to them. But by the very nature of the motives that drive them, they have greatly neglected the formations poor in metals. Thus in our region [around Paris] each kind of workman knows only the kind of quarry in which he works. He who searches for plaster does not know what is above and below the beds of gypsum; the quarryman does not know that he has the clay worker beneath him, etc.

Any man the least aware of the progress of the sciences will feel that a doctrine that would furnish, in relation to all the useful minerals, data similar to those that miners have on metallic veins, would be of the greatest

---

and distributions; "geognosy," if those directly observable features are extrapolated into a three-dimensional structural form; and "geology," if those structures are interpreted in turn as the products of causal processes. The previous sentence refers to the structural or *geognostic* level.]

6. ["Botany" is here used in a restricted sense to mean just the description and classification of plant species. The analogues of the inorganic field studies Cuvier had just mentioned would cover—in modern terms—ecology, biogeography, etc.; but these sciences had hardly begun to be developed in Cuvier's time.]

7. [The Marly machine was an elaborate series of hydraulic pumps, first installed in the late seventeenth century, which raised water from the Seine to serve Versailles on the plateau above. Shortly after Cuvier wrote this, it was in fact replaced by new pumps powered by steam.]

importance to society; and that if it were extended to all known minerals, it would form a branch of natural philosophy as beautiful as it would be curious.

It is probable that the surface of the globe, and the meager part of its interior that we are able to penetrate, would have been studied principally from this point of view, had one found nothing but inorganic minerals there. Since these minerals must have been deposited originally in some kind of order, one would not initially have expected to see in their disposition the proofs of successive actions and revolutions, had it not been that a very large part of their beds teemed with the debris of organisms.

In fact it is fossils and petrifactions that, by exciting curiosity and awakening the imagination, have made geology take too rapid a course, and have made it move too carelessly beyond the first bases that it should have founded on facts, carrying it in search of causes, which should only have been its final result. In a word, they have changed it from a science of facts and observations into a fruitless web of hypotheses and conjectures, so much at odds with one another that it has become almost impossible to mention its name without provoking laughter.

Fossils and petrifactions were at first considered to be sports of nature, without expanding much on what was meant by that. But when more careful study had shown that their general form, their inner texture, and in many cases their chemical composition were the same as those of the analogous parts of living organisms [corps vivans], it became necessary to admit that these objects, in their time, had also enjoyed life, and consequently that they had existed at the earth's surface or in the waters of the sea. How then were they found entombed in immense masses of stone and earth? How did these marine bodies find themselves transported to the summits of mountains? Above all, how had the order of climates been totally inverted, so that one found the products of the tropical zone near the pole?

Once it was finally seen that almost all the surface of the globe had been covered to an incalculable depth, it became essential to try to imagine what widespread and powerful causes had spread them thus. *Genesis*, and almost all the traditions of pagan peoples, offered one to which it was natural that physicists [physiciens] first had recourse. This was the Deluge. Petrifactions were taken to be the proof of it; and for almost a century works of geology only contained attempts to find the physical causes of this great catastrophe, or to deduce the present state of the surface of the globe as its effect. Their authors forgot that the Deluge is presented in *Genesis* as a miracle, or as an immediate act of the will of the Creator, and that consequently it is quite superfluous to look for its secondary causes.[8]

But toward the first third of the eighteenth century, it came to be thought that a single inundation, however violent, could not have produced such immense effects, the magnitude of which was accentuated daily. It was thus felt necessary to grant a long series of operations, whether slow or sudden; and those geologists who still accorded a real existence to the Deluge considered it simply as the last of the revolutions that have contributed to putting our globe into the state in which we see it.

Once this step was taken, hypotheses no longer knew any bounds. In this part of natural history one saw the rebirth of the systematic method of Descartes, which Newton seemed to have banished forever from the physical sciences. Each [author] devised some principle, found in advance a priori, or based only on a very small number of partial [i.e. local] observations, and used all the forces he could muster to submit to it—well or badly—all the facts made known to him. But in the midst of all these endeavors, by an almost inconceivable misfortune, the extension of factual knowledge was almost entirely neglected. And when one considers that Leibniz and Buffon are among the philosophers of whom I speak, it can well be admitted that it was not by lack of genius or talent that such a false route was taken.

Thus the number of geological systems has increased to the point that today there are more than eighty of them, and it has been necessary to classify them in a certain order, simply to help one to memorize their main features. [Even] the best example, put forward some thirty years ago by several savants, has so little inhibited additions to this long list, that we see new systems hatched every day, and the scientific journals are full of the attacks and defenses that their authors make against each other.[9]

How can so many men of spirit, full of science and good faith, be so little in accord, and continue so long with such controversies? The reason is very simple: if one of them was right, neither he nor the others could know it. To know whether a fact is due to a cause, it is necessary to know the nature of the cause and the circumstances of the fact. For what are the authors of geological systems, in the present state of the sciences, if not people who

8. [This had long been a standard formula—used for example by Buffon—to avoid the need to specify the kind of physical evidence that would be expected, or the kind of natural cause that might have been responsible, if the Deluge had indeed been a real historic event. The sentence was omitted from the version printed in the Institut's *Mémoires*, presumably because even such a perfunctory reference to the Creator was deemed inappropriate there.]

9. [These are probably allusions, respectively, to the massive review of geological "systems" compiled by de Lamétherie in the volumes of *Théorie de la terre* (1797); to the theory of a steadily cooling earth, popularized (though not invented) by Buffon in his "Époques de la Nature" (1778); and to the stream of theories published in the preceding years in the *Journal de physique* and other periodicals.]

look for the causes of facts that they do not know? Can one imagine a more chimerical goal?

We forget indeed that we are talking not only of the nature and arrangement of the interior of the globe, but [also] of that of its outermost skin. The research of miners, of Pallas, Saussure, Deluc, and Dolomieu, and of the Werner school, have given us valuable generalizations—although not yet beyond challenge—on the Primary rock-masses *[montagnes]*. But the Secondary formations *[terrains]*, which are the most awkward part of the problem, have scarcely been touched upon; the most crucial points, on which necessarily depend the side that one takes in relation to causes, are still in question.

We could cite a multitude of examples; but to be brief we shall restrict ourselves to one or two. Did organisms *[êtres organisés]* live in the places where their remains are found, or were they transported there? Are *all* these [fossil] beings still living, or have they been totally or partially destroyed?

Is it not clear that the system of causes to be conceived ought to differ as white from black, according to whether these questions are answered in the affirmative or not? However, no one is yet able to respond positively; and what is still more striking, almost no one has considered that it would be good to be able to answer them before constructing a system.

This is why some want billions of years for the formation of the Secondary formations, while others claim they were formed in one year, about five thousand years ago; and all the positions intermediate between these two extremes also have their defenders.[10]

There are already ten or twelve hypotheses [just] for the partial [i.e. local] explanation of the formation of the Paris Basin. Yet none of their authors knows that at Grignon, in one little corner of that basin only a few square fathoms in extent, there are six hundred unknown species of shells, for forty or fifty that can be recognized [alive]; this is a fact reported by Mr. de Lamarck, through research that required several years. Neither do any of them know that our [Parisian] plaster contains the bones of twelve or fifteen quadrupeds that resemble none that can be seen either here or elsewhere, another fact that has only been brought to light by ten years' work. You can judge what explanations must be like, that are conceived

---

10. [The "billions" *(milliards)* are American, not British; that is, hundreds of millions. That extreme was represented for example by Lamarck's *Hydrogéologie* (1802) and Hutton's *Theory of the earth* (1795), with their implicitly vast timescales for the shifting positions of continents and oceans. The other extreme was represented by those—they included no serious naturalists—who had revived the traditional brief timescale based on a literal interpretation of Genesis; among them was Chateaubriand, in his famous *Génie de Christianisme* (1802). As suggested in a different context (chapter 8), Cuvier was trying to establish a middle way between the two extremes, *both* of which he regarded as incompatible with the evidence.]

quietly in the study *[cabinet]* by persons to whom these two little circumstances of the phenomena were unknown![11]

So what ought learned bodies to do in order to procure, for such an interesting and useful science, the growth of which it is capable, by directing its course toward a goal that is real and attainable? They ought to maintain toward it the conduct they have maintained toward all the other sciences, ever since their foundation: they ought to encourage through their eulogies *[éloges]*[12] those who report positive facts, and to keep absolutely silent on the successive systems. Besides, the authors of the latter do enough of their own talking. It is curious to see them *all* on the lookout for the discoveries that observers make: prompt to seize them, to adapt them to their [own] ideas, or to use them to arm themselves against their adversaries. It seems that anatomists, zoologists, and mineralogists are mere maneuvers whose destiny is to provide materials for their fantastic constructions.

Happily for the example they set for those who might be tempted to follow in their footsteps, these castles in the air are evaporating like vain appearances, and the more solid edifice of facts and of induction is beginning to arise. The ground plan, so to say, has already been traced out. The good people of the end of the eighteenth century set out the questions; they already resolved some of them, and they pointed to the only way by which the others can be resolved. The series of problems has been proposed. Nothing more than an enlightened perseverance is needed to fill out the framework, the ensemble of which will constitute the science.

It is not useless to the purpose of our report to set out here, as examples, some of the main objects that seem to us to need thorough study, in order to make geology a science of facts. This must be done before one can try out its forces, with some hope of success, on the great problem of the causes that have brought our globe to its present state.

According to us it is necessary

1. To study whether the great division of major [mountain] chains into a central ridge and two sets of lateral ridges, as recognized by Pallas and

---

11. [This paragraph was added to the original manuscript text, perhaps to emphasize the point by referring to local examples. A quarry at Grignon, west of Versailles, was the best locality for the fossil shells that Lamarck had been describing in the *Annales du Muséum;* the other example refers of course to Cuvier's own research. Thus Cuvier deftly allied himself with that part of his senior colleague's research that he found acceptable and indeed valuable.]

12. [The reading of elaborate "éloges," assessing the scientific achievements of its recently deceased members, was a major feature of the life of the Institut; Cuvier himself was frequently responsible for composing and reading them. They were far more than mere obituaries, for they were used—as implied here—to indicate clearly the *kinds* of scientific work that were approved (or disapproved) by the most powerful scientific body in France.]

elaborated by Deluc, is constant; and to examine, as Mr. Ramond has done in the Pyrenees, the causes that sometimes conceal it.[13]

2. To examine whether there is also some constancy in the succession of the Secondary beds: whether one kind of rock is always below another, and reciprocally.

3. To make a similar study in relation to fossils. To determine the species that appear first and those that only come later; to know whether these two kinds of species are never found together, and whether there are some alternations in their return; that is to say, whether the first [kind] came back a second time, and whether the second [kind] had then disappeared.

4. To compare fossil with living species, with more rigor than it has been done hitherto; and to determine whether there is a correlation between the age *[ancienneté]* of beds and the resemblance or nonresemblance between fossil and living organisms.

5. To determine whether there is a constant climatic relation between fossils and those living organisms that most resemble them; to know, for example, whether they have migrated from north to south, from east to west, or whether there have been mixtures and radiations.

6. To determine which fossils have lived where they are found and which have been carried there; and whether in this respect there are constant rules in relation to beds, species, or climates.

7. To follow the different beds in detail throughout their extent, in their folding, their dip, their faults, and their indentations; thus to determine which regions belong to one and the same formation, and which others have been formed separately.

8. To follow horizontal beds, and those that are inclined in one or more directions, in order to determine whether there is some relation between the greater or lesser constancy of their horizontality, and their age or their nature.

9. To identify valleys in which the reentrant and projecting angles correspond,[14] and those in which they do not; also those in which the beds are the same on both sides, and those in which they are different; in order to find out whether these two circumstances are related to each other, and if

13. [The supposedly invariable structure of mountain chains, in which a granitic axis was flanked symmetrically by Secondary formations. The Pyrenees, in which some of the highest peaks (e.g. Mont Perdu) consisted of limestone with fossils, had seemed the most puzzling exception; but Ramond had shown that the structure was normal although the altitudes were not.]

14. [At this period the usual way of describing what was later interpreted as an incised meandering valley; many of the valleys *not* of this form were interpreted much later (after Cuvier's death) as glacial in origin.]

each of them separately is related to the nature and the age of the beds that compose the hills that border the valleys.

The clarification of all these points is indispensable, if one wants to make geology into a body of doctrine or a real science, independently of any desire that one might have to find an explanation of the facts. But it is quite clear that they are still more necessary in order to succeed with that explanation.

Now, we dare affirm that there is not one of them about which anything is absolutely certain. Almost all that has been said about them is vague to a greater or lesser extent. Most of those who have spoken of the matter have done so according to what suits their systems, much more than according to impartial observations. Fossils alone, considered in isolation, could still furnish several hardworking savants with material for thirty years' study; and their relation to the beds [couches] will require many more years of travel, excavation, and other arduous research.

What service would a body such as ours not render to the natural sciences, if it were to direct toward positive — but long and arduous — research, those people in whom the craving for knowledge, and the contagious example of so many men of merit, could [otherwise] lead toward systems that are as useless as they are easy to create and seductive to one's pride.

The work of Mr. André, examined along these lines, offers us two quite distinct parts, of which only the first seems to us to fall within the competence of the Class. This is the part in which the savant reports on the observations he made during his travels. Faithful to the rules of the religious order to which he belonged, Mr. André followed, on foot, some quite numerous and extensive routes; he followed them as an enlightened [éclairé] observer,[15] and noted with care the elevations and depressions of the terrain, the nature of the rocks, and their disposition in relation to each other and to the horizon. As a model he took the geologist who most deserved that honor, the celebrated Saussure; that is to say, he described, in an absolute manner, each of the objects that struck him on the way, in the order in which he encountered them.

A [mountain] range traversed thus in several directions, and described with such care, becomes the subject of a general picture, which Mr. André has not failed to trace out. It is thus that he informs us about the part of the Alps that he has seen, which makes up the space between the Saint Gothard and the Little Saint Bernard [passes]. He then moves to the Jura, a Secondary range very different from the Alps, which he examined from the outflow of the Rhône up to the Rhine, that is to say, almost throughout its

15. [The former member of a religious order is here acknowledged as having carried out his fieldwork in the open-minded spirit of the Enlightenment; and, as one under vows of poverty, without benefit of the horses and carriages used by wealthy secular naturalists.]

whole length. The Vosges are a third range, of which Mr. André examined the part that extends from Épinal to Giromagny, and from Giromagny directly across to the Grand-Donon. Finally he describes the watershed *[crête de séparation]* from which the waters flow on the one side to the [Atlantic] Ocean, and on the other to the Mediterranean. He followed it from the Haut de Salins, near the Marche, to just past Cluny. He has also observed and described part of the plains that link the Alps to the Jura, and those that, beginning at the Saône, follow the course of the Rhine to Strasbourg.

Although throughout this [first] part of his work Mr. André alludes continually to the opinions that he seeks to prove in the second, the first is no less precious because of the large number of interesting facts that he describes there, and which are independent of any system. Such, first, are the cirques or circular spaces sunk between high and abrupt peaks, which he observed frequently in the Alps. Such also are his remarks on certain isolated [rock] pyramids: although they consist of diverse beds, and although the whole area around them must have been elevated by some kind of cause, their debris is not found at the foot of the pyramids.

In the Valais, Mr. André describes many escarpments and aqueous erosions that escaped Saussure, because the latter saw the lower part of the country, between Martigny and Brig, only for two days and always from the main road.[16] Mr. André also shows that this large valley, far from having projecting and reentrant angles that correspond on the two sides, widens and narrows alternately up to five times. The section on the Valais is in general one of the most complete in the work, Mr. André having traversed it a great number of times and by different routes.

In several regions of the Alps he points out examples of shaly beds that are contorted or curved in many directions, and that are difficult to reconcile with the usual theories. In general he seems to find little to favor the idea of the displacement of beds.

His description of Mont Blanc, which has much precision and clarity, makes one read with interest, even after that of Saussure (to the veracity and exactness of which he does perfect justice).[17] He describes Saint Gothard and its environs with the same care. He notes that the highest peaks are not in the central range; he observed a similar fact in the Vosges. This is the same as Mr. Ramond reported in the Pyrenees.

In his description of the Jura, he carefully distinguishes the compact limestone without petrifactions, which forms the central part of the range,

16. [This final phrase was omitted from the text published by the Institut, perhaps because such a suggestion of cursory observation impugned Saussure's towering reputation!]

17. [Saussure had not only described the region around Mont Blanc in detail, but had also been (in 1787) the first savant to set foot on the summit of the highest mountain in Europe.]

from the shelly limestones that make up the lateral and lower parts. He shows that some rolled cobbles [cailloux roulés] and big limestone blocks rounded by transport are like those of granite in the Alps. But there are also some of the latter in the Jura, although Saussure, who did not traverse it sufficiently, did not believe that at all. Mr. André cites several of them.[18] He talks of numerous caves and other degradations in this range. He describes the ice caves, and above all the one in the limestone five leagues from Besançon, of which he gives the temperature taken at different times of year, in order to show that it is very far from being the opposite of that of Dehon, as some have claimed.

His comparison of the Alps, the Jura, and the Vosges is interesting [curieuse]. In the Alps, there are longitudinal and transverse valleys; in the Jura they are almost all longitudinal; in the Vosges, almost all oblique. It is known that the Pyrenees have yet a fourth structure, and that the valleys there are nearly all perpendicular.[19] The Vosges are unique in the quantity of sandstone and conglomerates [poudingues] that covers their isolated summits, and which appear to be the remains of an immense plateau.

These details show that Mr. André has observed with care the countries he has traversed, and that the facts he has recorded in his works can be very valuable for positive geology, at least those that concern the mineral masses. We consider that, although he has not been concerned at all with fossils, he will take in this respect a distinguished position among observational geologists [observateurs géologistes].[20]

To his own descriptions of the countries he has seen, he adds several about those where he has never been, drawn from the best authors, such as Messrs. Saussure, Deluc, Dolomieu, Ramond, and Patrin. These extracts should not be summarized a second time. We shall confine ourselves to saying that the author notes that there ought to be much analogy between widely separated regions, and that the theories that are applicable to our countries should [also] be so, more or less, to the whole earth.

At the end he does say a few words about fossils, but only drawn from other naturalists.

After having thus established his facts with much care, from his own

18. [These large "erratic blocks" of granite on the flanks of the Jura hills, clearly matching the granites in situ in the Alps on the far side of the Swiss plain, were often invoked as some of the best evidence for an exceptional "catastrophic" event. They remained one of the most puzzling of all geological phenomena, until the development—some three decades later—of theories of the formerly vast extension of Alpine glaciers during some kind of "Ice Age."]

19. [All these terms refer to the dominant orientation of the valleys in relation to that of the central ridge of the range concerned, as seen in plan view on a topographical map.]

20. [The phrase indicates the approval that Cuvier was now prepared to bestow on "geology" and its practitioners: a "geologist" could also be a good "observer" *in the field*.]

work or from the most respectable authorities, Mr. André turns to the consequences that he believes follow from these different facts. After all that we have said at the start of this report, it should not be expected that we shall pass any judgment on this part of his work. But we are not prohibited from giving some idea of it. He thinks that the present arrangement of the surface of the earth dates from a moderately remote epoch; and like Mr. Deluc and Mr. Dolomieu he seeks to prove this from the course of collapsed and eroded structures [éboulemens, atterrissemens]. He thinks furthermore that this arrangement is due wholly to a cause that was unique, general, uniform, violent, and sudden. He seems to attribute even the transport of exotic [étrangers] fossils to this same cause. He seeks to show that neither volcanos nor earthquakes, neither rivers nor [marine] currents, could have formed the surface of the earth as it is today.

Several celebrated naturalists also share these ideas, particularly if they are restricted to the last [i.e. most recent] change. Your commissioners even think they are personally able to adopt them in part, although they can very well conceive that the reasons that persuade them may not have the same influence on everyone.[21] For the reasons they have expounded earlier, however, they do not believe themselves in duty bound to commit the Class to pronouncing on such subjects. But they do not hesitate at all in proposing that it should acknowledge the esteem that is due to Mr. André for his painstaking research, and for the enlightened zeal that has led him to continue his useful work at such an advanced age.[22] They have not the slightest doubt that the work of this respected savant will be welcomed by naturalists as befits a compilation so rich in interesting facts.

Presented at the Imperial Palace of the Louvre, on 11 August 1806; *signed*, Le Lièvre, Haüy, and Cuvier *(rapporteur)*. The Class approves this report, and adopts its conclusions. [This copy] certified to agree with the original.

G. Cuvier
Permanent Secretary

Translated from Cuvier, Haüy, and Lelièvre, "Rapport sur l'ouvrage de M. André" (Report on Mr. André's work, 1807).

21. [This clause was added to the original draft; it may indicate that Cuvier and Haüy (and perhaps Lelièvre too) were more inclined to see a core of historicity in the ancient traditions of a Deluge—including the biblical story—than were many in their audience at the Institut. The latter part of the previous sentence, confining the plausibility of the suggestion to the most recent "catastrophe," was also an addition.]

22. [André was nearly seventy-eight when the report was read, whereas Cuvier was just short of thirty-seven; André died less than a year later.]

THE PROGRESS OF

GEOLOGICAL SCIENCE

———

Cuvier's new appreciation of the value of geological work was proba-bly based not only on what he saw it could do for his own research on fossil vertebrates, but also on what he learned about geology around this time, in the course of preparing a review of *all* the natural history sci-ences. Early in 1808, the Institut presented Napoleon with a massive compilative review of the progress of the sciences since the start of the Revolution in 1789.

Cuvier shared the burden of preparing the report with the other permanent secretary of the scientific class of the Institut. Jean-Baptiste-Joseph Delambre (1749–1822) dealt with the "mathematical" sciences; Cuvier, with the "natural" sciences of chemistry and natural history and with the applied sciences (fig. 15).[1] Even that division of labor left each with a vast field to cover, although of course they made full use of their colleagues as informants on particular subjects. The arrangement of their report reflected the classification of the sciences that was standard at this

1. A similar report on the humanities was presented by the historian Bon-Josephe Dacier (1742–1833) two weeks later (*Rapport historique,* 1810); it reviewed the progress of research in the realm of the class for history and ancient literature. The class for "moral and political sciences," or in modern terms the social sciences, had been suppressed by Napoleon in 1803; like certain authoritar-ian politicians in modern times, he regarded such studies as potentially subversive.

RAPPORT HISTORIQUE

SUR LES PROGRÈS

DES SCIENCES NATURELLES

DEPUIS 1789,

ET SUR LEUR ÉTAT ACTUEL,

Présenté à SA MAJESTÉ L'EMPEREUR ET ROI, en
son Conseil d'état, le 6 Février 1808, par la Classe
des Sciences physiques et mathématiques de l'Institut,
conformément à l'arrêté du Gouvernement du 13 Ventôse
an X;

*RÉDIGÉ par M. CUVIER, Secrétaire perpétuel de la
Classe pour les Sciences physiques.*

IMPRIMÉ PAR ORDRE DE SA MAJESTÉ.

A PARIS,

DE L'IMPRIMERIE IMPÉRIALE.

M. DCCC. X.

FIGURE 15 The title page of Cuvier's *Rapport historique* (1810) on the progress of the "natural sciences" since 1789; the title page of Delambre's report on the "mathematical sciences" was otherwise identical. As the design suggests, the reports were intended to serve the nationalistic cultural politics of Napoleon's imperial regime, but in practice both savants reviewed their sciences from a highly international perspective.

period; its preparation must have made them, at least at this moment in history, two of the most widely knowledgeable men of science in the world.

Cuvier's primary divisions of "natural history" followed implicitly the ancient—and still convenient—concept of four primary "elements":

there were major sections on meteorology ("air"), hydrology ("water") and the natural history of minerals ("earth"); the fourth and longest section dealt with the natural history of organisms *(corps vivans)* ("fire," in its traditional equation with life).

The section on minerals was subdivided into "mineralogy proper" (in essentially its modern sense) and "geology": in Cuvier's work the latter term had at last come in from the cold. His section on geology (text 14) is a valuable summary of the state of the science at this time, as seen by a savant of broad interests and highly cosmopolitan outlook. Although the whole report was intended to bring cultural glory to the French Empire, which Napoleon was currently extending by conquest into most of western Europe, its authors explicitly disdained any narrow chauvinism, and made their survey of the sciences fully international.[2]

Cuvier presented geology as a science of very recent origin, dating only from the thorough researches of naturalists such as Pallas, Saussure, Deluc, and Werner. Their fieldwork, above all, had provided a descriptive "physical geography" or "positive geology" of specific regions, which was a foundation on which a more explanatory "general geology" might in due course be built. The great sequences of rock formations, the vast scale of their subsequent erosion, the huge masses of gravel deposits, and the widespread relics of volcanos where none are now active: all these were presented, as before, as the "infallible marks of great revolutions" in the form of the earth's surface. But there was no particular stress on the sudden character of most of these major changes. As usual, however, Cuvier did propose that a marine inundation—and implicitly a sudden one—had caused the extinction of terrestrial vertebrates; and, following Deluc and Dolomieu, he cited the recent "alluvial" deposits as important evidence for the geologically recent date of that last revolution. Unsurprisingly, fossils received in general much more attention from Cuvier than they might have had at the hands of most other authors at this time. Specifically, he described how the various formations are characterized by particular fossils or groups of fossils: this kind of general correlation between rocks and fossils was clearly well understood.

Toward the end of his review of geology, Cuvier considered more briefly the conjectures that could be made about the *causes* of all these

---

2. Cuvier's very sparse citation of sources in English is unlikely to reflect either wartime politics or plain ignorance. In the eyes of savants on the continent of Europe, Britain at this time was, in terms of the sciences of the earth, little more than an offshore island. North America, like Siberia and India, was almost as it were in outer space: non-European lands were of great importance for the natural history sciences, but only as collecting grounds for specimens and observations, not as sources of explanatory or theoretical ideas.

dramatic effects. As usual, he reviewed the proliferation of "systems" with evident scorn; he argued that causal explanations were still premature, as long as there was so little reliable factual evidence on which to base them. But he now showed some ambivalence about "systems": in general he claimed that they tended to inhibit progress in geology, by discouraging the search for reliable facts; but he also conceded, at least in passing, that they might encourage that search.

TEXT 14

—

[Historical Report on the Progress of Geology since 1789, and on Its Present State]

GEOLOGY

The formation and ordering of this great catalogue of minerals, and even the most complete description of the properties of each, is still only a part of their [natural] history.[3] It is necessary to add to it the knowledge of their respective positions, and of their distribution in all the beds of the globe that we can penetrate.

PARTICULAR GEOLOGY

That is the object of positive geology and physical geography; it is a kind of particular geology, which is the basis for general geology. In it one examines in depth the mineral structure of a given region, and the nature of the rocks or other minerals that form its mountains, hills, and plains, as well as their relative positions. It is, so to speak, a wholly modern science. Pallas has given fine examples of it for Russia,[4] Saussure for the Alps,[5] and Mr. Deluc for certain regions of Holland and Westphalia.[6] The school of Werner has made in this respect the finest researches in Saxony and in

3. [The allusion, carried over from the previous section of the *Rapport*—on mineralogy—is to footnotes that refer to works such as Haüy's *Traité de minéralogie* (1801) and Brongniart's *Traité élémentaire de minéralogie* (1807).]

4. In his "Observations on the formation of mountains," *Academy of Petersburg, 1777* [1778], and in his *Travels* [*Reise*, 1771–76].

5. *Travels in the Alps*, Neuchâtel, 1776–96, 4 vols., 4to.

6. *Letters to the Queen of England on the history of the earth and of man* [i.e. *Lettres physiques et morales*], The Hague, 1778, 6 vols., 8vo.

several other regions of Germany and neighboring countries.[7] Mining areas have been examined, as one would expect, with even more care than others: immediate [economic] interests demanded it, and Saxony and Hungary, where the art of mining has been practiced since time immemorial, have had the finest [natural] historians.

GEOLOGY OF FRANCE

The physical geography of France has been cultivated in recent times with no less ardor than that of foreign countries. The courses of Rouelle, those of Valmont de Bomare, of Daubenton, and of Mr. Sage, as well as their elementary works, have begun to spread in our nation a taste for mineralogy, long concentrated in Germany and Sweden. Museums [cabinets] have been formed in our main towns, and mineralogical travels undertaken in almost all our provinces. Even before the period on which we are reporting, Gensanne and Soulavie had described Languedoc, Besson the Vosges; our iron mines, the principal riches of this kind in France, had been examined by Dietrich,[8] and Picot-la-Peyrouse had described those of the county of Foix;[9] Palassou, and more recently Mr. Ramond, had published in detail on the Pyrenees.[10]

The Council of Mines, founded in 1793, when the interruption of all contact with foreign countries made us feel the need to exploit our own territory, has given a quite new impulsion to this kind of research. The engineers sent by its orders into the various departments have studied their mineralogy; and a fairly large number of exact descriptions—made above all by Mr. Dolomieu, Mr. de Gensanne, Mr. Lefebvre, Mr. Duhamel the younger, Mr. Baillet du Belloy, Mr. Héron de Villefosse, Mr. Cordier, Mr. Rosière, and Mr. Hericart de Thury—have already been collected in the *Journal of mines*.[11] Our coal mines have excited lively attention, and

7. The particular geological works from the school of Werner are as numerous as they are important; the most complete exposition of their results yet is found in the *Geognosy [Lehrbuch der Geognosie]* of Reuss, Leipzig, 1805, 2 vols., 8vo, in German. Among their number are those of Messrs. von Buch, Sturl, Leonhard, Lazius, Noze, Voigt, Freiesleben, Wrede, etc. We need not cite the most celebrated of Werner's students, the illustrious and courageous Mr. von Humboldt. It is also worth consulting the older works of Charpentier, von Born, etc.

8. *Description of the mineral deposits, forges, and saltworks of the Pyrenees [Description des gîtes de minerai, des forges, et des salines des Pyrénées]*, by Baron de Dietrich, Paris, 1786, 4 vols., 8vo.

9. *Treatise on the iron mines and forges of the county of Foix [Traité sur les mines de fer et des forges du comté de Foix]*, by [Picot] de la Peyrouse, Toulouse, 1786, 1 vol., 8vo.

10. [Palassou,] *Essay on the mineralogy of the Pyrenees [Essai sur la minéralogie des Pyrénées]*, Paris, 1781 [1784]; *Observations made in the Pyrenees [Observations faites dans les Pyrénées]*, by Ramond, Paris, 1789, 1 vol., 8vo.

11. This collection began in Vendémiaire, Year III [September–October 1794], and continues successfully. Germany has several analogues [i.e. periodicals], such as those of Mr. von Moll, Mr. von Hoff, etc.

Mr. Duhamel Sr., Mr. Lefebvre, Mr. Gillet-Laumont, and Mr. de Gensanne have successfully been concerned with their outcrops, their folds, the faults and stony veins that interrupt them, and all the details of their exploitation and use. The rich mines that by force of arms have fallen to the power of France in the conquered departments[12] have been carefully examined and described, and have enriched science as well as the empire. In the old provinces,[13] various mines of metals of practical value have been discovered or described, ranging from mercury and copper to chromium and manganese; and numerous quarries of stones appropriate for all kinds of construction, from the marbles and porphyries that enrich our palaces, to the refractory bricks from which kilns are made. Amid all this research, a mass of minerals is encountered, which, without yet having any immediate use, nonetheless belongs to the great system of our physical geography, and furnishes precious materials for chemical research.

Thus emeralds have been found near Limoges, by Mr. Lelièvre; pinite at Le Puy, by Mr. Cocq; native and oxidized antimony in Germany, by Mr. Schreiber; uranium oxide at Sémur, by Mr. Champeux, and at Chanteloup near Limoges. One of the most interesting of these discoveries is that of an iron chromate mine, made in the department of the Var by Mr. Pontier, and of which we spoke only a moment ago.[14]

GENERAL GEOLOGY

These mineralogical descriptions of the various regions, assembled and compared, offer many points of conformity which, by that very conformity, must be due essentially to the structure of the crust of the globe. The series of these common results, which are found more or less the same everywhere on earth, is what properly constitutes the science of positive or general geology. This, in assigning the laws of the respective positions of the various minerals, is of the highest importance in guiding the search for them.

As usual, it was [economic] interest that furnished the first outlines of the picture. Mountains rich in metallic veins were the first to be studied, and they were distinguished from those in which the horizontal beds are most often poor in metals. That point was reached toward the middle of the eighteenth century. Soon it was noticed that the rocks with veins are always

12. [In particular, the rich coalfields of the Rhineland and the future Belgium had come under French control—the latter within new administrative departments—through Napoleon's campaigns and the earlier revolutionary wars, greatly enlarging the modest coal reserves within the frontiers of France itself.]

13. [The administrative regions into which France was divided before the Revolution, when they were replaced by more numerous and smaller departments.]

14. These memoirs and many others can be found in the *Journal of mines* [The reference to Pontier's discovery was in Cuvier's preceding survey of mineralogy.]

close to the still more compact rocks that form the highest mountain chains; that both are lacking in the remains of organisms that fill the ordinary beds; and finally, that the latter, situated on the flanks of the former, must have been formed after them.

Hence came the distinction, fundamental in geology, between the Primitive formations *[terrains primitifs]* that are considered to predate organized life, and the Secondary formations *[terrains secondaires]* deposited on them by water, which swarm with the remains of organic productions. It seems that Lehmann and Rouelle were the first clearly to classify the formations along these lines.[15] But many developments of these ideas still remained to be given: the Primitive formations themselves are of several kinds, and probably of several ages; and still less can one fail to recognize a long succession among the Secondaries.

Granite and similar rocks form the massif that bears all the other formations, and that pierces them, rising as pinnacles, ridges, or plateaus along the axes of the highest chains. On their flanks are beds of gneiss, schist, and other foliated rocks, the normal matrix of metallic veins; and mixed with them or resting on them are various granular marbles. The beds of all these rocks are broken, tilted, and disordered in a thousand ways. This is what Mr. Pallas has reported for the mountains of Russia; what Mr. de Saussure and Mr. Dolomieu have confirmed for those of Europe, and on which Mr. Deluc has expanded. The Pyrenees seemed to be an exception to the rule; but Mr. Ramond has shown that this exception is only apparent, and due solely to the fact that the schists and limestones on the Spanish side are higher than the central granitic axis.[16]

Mr. Werner and his students have given many more details on the superposition of these Primitive formations; but they have perhaps subdivided them too much for their observations to be applicable in their entirety to other countries than those in which they were observed. In his *Theory of veins*, Mr. Werner has also given an interesting collection of observations on the course of these singular fissures, and has sought to determine in a precise manner the age of the metals, from the way in which the veins cut each other; for if, as it appears, veins are simply cracks filled subsequently, those that traverse others must be later than them.[17]

---

15. On the history of geology, principally in the eighteenth century, various articles of the *Dictionary of Physical Geography* by Mr. Desmarest [*Géographie physique*, vol. 1, 1794–95], in the *Methodical Encyclopedia [Encyclopédie méthodique]*, may be consulted.

16. *Journey to Mont Perdu [Voyage au Mont-Perdu]*, Paris, 1801, 8vo.

17. *New theory of the formation of veins [Nouvelle théorie de la formation des filons, etc.]*, translated from the German by Mr. Daubuisson, Paris, 1802. [Mineral veins were a contentious issue in the de-

The Secondary formations are less easy to observe than the Primitive: more generally horizontal, it is rarer to find substantial vertical sections of them; and their various arrangements do not have anything like as much uniformity. In what is known of them, however, a certain order of superposition can be noticed. The hard limestones filled with ammonites, the shales and coal seams marked with imprints of ferns or palms, the chalk full of flints molded into sea urchins and crystalline belemnites, the coarse limestones composed of shells more like those of our present seas: these follow each other according to certain laws. Clays, sands, and gypsum cover them here and there, and contain in confusion rolled shells, the bones of quadrupeds, and the imprints of fish.

These immense deposits, furrowed by streams and rivers, interrupted by flows of lava and other volcanic products, filled up or bordered by alluvial deposits, covered in many areas by an abundance of rolled cobbles, bearing here and there the evident debris of older strata, the infallible marks of great revolutions: these constitute the greater part of our continents.

In this great ensemble a mass of details attracts the notice and the reflections of the observer. Enormous blocks of Primitive rocks such as granite are scattered on the Secondary formations, as if they had been dropped there, and seem to indicate great eruptions. Mr. Deluc has stressed this fact; Mr. von Buch recently observed that the blocks of north Germany resemble the rocks of Sweden and Lapland, and seem to come from that region.[18] Masses of rolled cobbles occupy the mouths of large valleys, and seem to indicate large debacles; Mr. de Saussure has carefully described several examples of this. Sometimes the beds of these cobbles cemented into conglomerates [poudingues] are elevated, which proves that there were upheavals [bouleversemens] after some of these debacles. Examples are found even in Siberia, where Mr. Patrin has described them; Mr. von Humboldt has found them in abundance in the vast plain watered by the Amazon.

In general the Secondary formations, which one must suppose were formed peacefully and by way of deposition or precipitation, have not all retained their original position, for some are seen tilted, set on edge, faulted,

---

bates between "Neptunists" and "Plutonists": Werner argued that veins had been filled from above,i.e. from the overlying proto-ocean, not by injection from below; but Cuvier's point about their relative dating was valid either way.]

18. [As mentioned earlier, such blocks were (and still are) termed "erratics," for just that reason. Their attribution to transport by drifting icebergs, or still more radically by vast ice sheets, was not proposed until many years later, and not generally accepted by geologists until long after Cuvier's death.]

and upheaved *[bouleversés]*. Mr. Deluc also has the merit of having shown all these disturbances *[désordres]* very well.[19]

Volcanos are a cause of change that is still active at certain points on the surface of the globe. It has been interesting to study their mode of action, the nature and characteristics of their products, the degree of heat with which those products leave the crater; and indeed to seek to conjecture the depth of the source *[foyer]* from which they emanate, the causes that occasion and sustain their burning, and those than maintain the fusion of the lavas.

Dolomieu[20] and Spallanzani are those who, in recent times, have undertaken this kind of research with the greatest perseverance. They have both collected and described the products of Vesuvius and Etna with great care. Mr. von Humboldt, on his return from climbing the higher peaks and still more terrible volcanos that serrate the Cordillera of the Andes, had the good fortune to see the most recent eruption of Vesuvius close at hand. The volcano of the island of Réunion has yielded precious observations to Mr. Huber and Mr. Bory Saint-Vincent.

One of the most remarkable facts that seem to have been established is that the fire of volcanos does not have anything like the high degree of heat that had been attributed to it. Dolomieu assured himself of this, when examining the action of the lava on the various objects that it enveloped in 1798, in a village at the foot of Vesuvius. In that way he explained how it could incorporate the highly fusible crystals that it often contains, without melting them. Nonetheless, the lava is very fluid; it insinuates itself into the smallest interstices of objects. On the Isle of Bourbon there are trunks of palm trees in which all the cracks are filled with it (this is one of Mr. Huber's observations). When it flows, it effervesces and gives off thick vapors. Does it only catch fire on contact with the atmosphere, and does it allow

19. The letters of Mr. Deluc to Mr. de Lamétherie, collected in the *Journal of physics* for the years 1789–91, and the same author's *Geological letters to Mr. Blumenbach [Lettres sur l'histoire physique de la terre]*, Paris, 1798, 8vo, contain the exposition of his detailed ideas on the theory of the earth. [The former collection was that on which Cuvier had reported to his German friends, while still in Normandy (text 2).]

20. *Lipari Islands [Voyage aux Iles de Lipari]*, 1783; *Journey to the Iles Ponces* and *Catalog of the products of Etna* [i.e. *Mémoire sur les Iles Ponces*], 1788; and above all his last memoirs in the *Journal of physics* and *Journal of mines*. Add to these works the memoirs of Mr. Fleuriau de Bellevue, those of Mr. Daubuisson, and the *Essay on the volcanos of Auvergne [Essai sur les volcans d'Auvergne*, 1789] by Mr. de Montlosier.

some substance to escape, which maintains its fluidity at this moderate degree of heat, as Kirwan and Dolomieu suspected?[21]

The quantity of these lavas is enormous. The Delucs tried to show that the whole mass of volcanic mountains is formed from the very products of their eruptions; and the number of volcanos was once much more considerable than it is today. This is what was recognized, as soon as there were sufficient notions about modern lavas to be able to compare them with ancient ones. Mr. Desmarest was one of the first to concern himself with this kind of research. Above all, he made known the extinct volcanos of Auvergne. He climbed to their craters; he followed the trails of their lavas; he has seen them cracked into basaltic columns; and it was from his observations that all basalts, rocks fairly similar to certain lavas, were long assigned to a volcanic origin. Mr. Faujas has done similar work on the extinct volcanos of the Vivarais;[22] Fortis, on those of the Vicentino,[23] etc.

However, it seems that formations that resemble lavas do not all have the same origin. Such are the rocks termed "wackes."[24] They are of great extent in certain regions of Germany. They are indeed horizontal, they do not reach into any hill that could be considered a crater, and they often rest on highly combustible coals, which they have not altered at all. They are thus not volcanic. Mr. Werner has demonstrated these facts very well, and as a result of his observations a multitude of formations have been stripped of the [volcanic] origin to which they had been attributed. (Even so, it remained the opinion of Hutton and Mr. James Hall that they had melted in situ, at the time of a general and violent heating suffered by the globe.) The resemblance of the stone is thus no longer sufficient to make one believe in an extinct volcano; traces of an eruption are still needed. But when these traces are clear, one cannot refuse to concede the point. Besides, Mr. von Buch and Mr. Daubuisson, the distinguished students of Mr. Werner, have acknowledged the volcanic nature of the peaks in Auvergne.

By examining in this way the various regions of the globe, one finds that

21. [The Irish chemist Richard Kirwan (1733–1812), a vigorous critic of Hutton, is one of the few anglophone savants mentioned by name in Cuvier's report.]

22. *Researches on the extinct volcanos of the Vivarais and the Velay [Recherches sur les volcans éteints du Vivarais et du Velay]*, Paris, 1778, folio; *Mineralogy of volcanos [Minéralogie des volcans]*, Paris, [1784], 8vo.

23. *Memoirs on the natural history of Italy, and principally on its oryctography [Mémoires sur l'histoire naturelle, et principalement sur l'oryctographie de l'Italie]*, Paris, 1802, 2 vols., 8vo.

24. [The word is given here in its original German form; Cuvier used a French spelling *(vake)*; it survives in the modern geological term "greywacke." To the naked eye—before the development of petrological microscopes and thin-section techniques—these dark fine-grained rocks (now attributed to a sedimentary origin) were almost indistinguishable from basalts of volcanic origin.]

volcanos were once infinitely more numerous than they are today. They occur the whole length of Italy, and the seven hills of Rome are the remains of a crater, according to Mr. Breislak.[25] The banks of the Rhine bristle with them; they are seen in Hungary, in Transylvania, and at the far end of Scotland.

The observation of extinct volcanos has even thrown light on the nature of volcanos in general. Thus Dolomieu, in studying those of Auvergne, believed he had perceived that their source must be beneath an immense plateau of granite, which is now covered by the products of their eruptions. It is thus that the otherwise unknown stones that so many lavas contain are explained; it is, however, not entirely proven that some of them have not been crystallized while the lava was still liquid.

For the rest, whatever the number of ancient volcanos may have been, it is not they that disrupt the other beds. It seems to be well proven, from the Delucs' observations, that they can only have a local influence, in piercing these beds and covering them with their products.

The great age of some of them [ancient volcanos] is shown by the marine beds that have been formed on, or alternate with, their lavas. But how can the fire of volcanos be sustained at these inaccessible depths? Why are almost all active *[brûlans]* volcanos so close to the sea? Is saltwater necessary to these internal reactions *[fermentations]*? Is this the origin of the saline products that accumulate on the edges of craters, and of which some are still found in extinct volcanos, as Mr. Vauquelin has noticed in Auvergne? Here are questions that could occupy physicists *[physiciens]* for a long time to come.

ALLUVIA

Running waters are another cause of change, less violent but today more widespread than volcanos. They carry stones, sand, and earth from elevated places, and deposit them in low areas, when they lose their rapidity. Hence the alluvia by the banks of rivers, and above all at their mouths; it is thus that the [Nile] Delta of Egypt was formed and is still growing. Lower Lombardy, and a part of Holland—Zealand—have no other origin.[26] The soils thus formed are the most fertile in the world. But the inundations that create them also devastate them from time to time; and if they are surrounded too soon by dikes, they are exposed too far below the level of

---

25. *Travels in Campania [Voyages dans la Campanie]*, Paris, 1801, 2 vols., 8vo.

26. [The plain and delta of the river Po, in northern Italy; and the southern coastal province of the Netherlands, which Cuvier, like many non-Dutch authors, refers to as Holland.]

the river. This is the case in Holland, which, in many areas, can only be drained by the power of machines. The most pressing [economic] interest thus demanded that this branch of geology be studied, in order to find both the means to profit from these new lands and the way to prevent their inconveniences.[27]

[Natural] philosophers studied them for another reason: they believed they could find there the surest index of the epoch at which our continents suffered their last revolution. In effect, these alluvia accumulate fairly rapidly; and since, at their origin, they must have done so still more quickly, their present extent seems to accord with all historical monuments, in making us regard this revolution as fairly recent. Mr. Deluc and Mr. Dolomieu are those who still seem to us to have best developed these kinds of facts.

FOSSILS

But the most stimulating of all that these geological studies offer is, beyond contradiction, what concerns the innumerable remains of organisms with which the Secondary formations swarm, and of which in some places they even seem to be wholly composed.

It was long since noticed that productions of the sea thus cover the solid ground with their masses, up to altitudes infinitely higher than those attained nowadays by the most terrible floods. A more careful examination had made it known that the productions that cover each region are almost never those of the neighboring seas, and even that a large number of them have never yet been recovered from any sea. The same observation applied to the debris of plants and to the bones of terrestrial animals.

Such a great spur to curiosity has produced its effect. Fossils, petrifactions, have been collected from every part; and descriptions of them begin to form a large and wholly specific series, which adds many species to those of beings known alive. At present Mr. de Lamarck is the one who has concerned himself with fossil shells the most fruitfully and with the greatest perseverance; he has made known several hundred new species, from the Paris region alone.[28] The fossil fish of the Verona region have been described and engraved magnificently through the care of Mr. de Gazola.[29]

---

27. [The lands were not only new geologically but also—to France—new politically and economically: Napoleon had annexed the Netherlands and set up a puppet state in Lombardy.]

28. In the different volumes of the *Annals of the Museum of Natural History* [1802–9].

29. *Veronese fossil ichthyology [Ittiolitologia veronese]*, folio [1796]. Only a slender part has yet appeared, although all the plates are ready.

Fossil plants have been less studied. There are some in recent beds quite similar to those of today. Mr. Faujas has described several of them; but the coal seams and shales conceal unknown ones. Count Sternberg has recently given an essay on them;[30] they are also beginning to be collected and engraved in England and Germany. In the latter country the work of Mr. de Schlotheim may be cited.[31]

Among these astounding monuments of the revolutions of the globe, none should have inspired more hope of more illuminating information than the remains of quadrupeds, because it was easier to determine their species, and the resemblances or differences that they might have with those that live today. But as their bones are almost always found scattered, and most often damaged, it was necessary to devise a method of recognizing each bone, each portion of bone, and to relate them to their species. We shall see elsewhere[32] how Mr. Cuvier has achieved that. He examined the bones in question according to this method, and he thus re-created several large species of quadrupeds, of which no individual is still living on the earth's surface. The gypsum quarries around Paris have alone furnished him with more than ten, which even form new genera. More recent formations have the bones of known genera, but of species that are not known at all. It is only in the alluvia and other deposits that are still being formed daily that one finds the bones of our present species.[33]

The unknown bones are almost always covered by beds full of seashells. It is thus some marine inundation that annihilated the species; but the influence of this revolution, by its very nature, was not perhaps exercised on all marine animals. However, it is beyond doubt that the deepest and consequently oldest of the Secondary strata swarm with the shells of other organisms [productions] that it has hitherto been impossible to discover in any of the oceanic regions. Since species similar to those that are fished for today exist only in the superficial beds, one is authorized to believe that there has been a certain succession in the forms of living beings.

Coals also seem to be ancient products of life: they are probably the remains of the forests of those remote times, which nature seems to have put

30. It is again in the *Annals of the Museum* that Messrs. Faujas and Sternberg have published their papers.

31. [Cuvier's allusions are probably to Parkinson, *Organic remains* (1804–11), of which the first volume (1804) was devoted to fossil plants; and Schlotheim, *Flora der Vorwelt* (Flora of the former world, 1804).]

32. [In the following section of the *Rapport historique* (not translated here), on comparative anatomy.]

33. Mr. Cuvier's papers on the reconstruction [*réintégration*] of the extinct [*perdues*] species of quadrupeds are still only in the *Annals of the Museum of Natural History* [1804–10].

into reserve for the present age.[34] More useful than any other fossil,[35] they necessarily attracted attention early on. Their depth and the nature of the stony beds that enclose them indicate their antiquity; and the wholly strange species of plants that they conceal accord with the animal fossils, in proving the variations that organized life [organisation] has undergone on earth.

Even the yellow amber conceals unknown insects and sometimes is found in equally unknown fossil woods.

In view of such an imposing—even terrible—spectacle as that of these remains of life, forming almost all the ground on which we tread, it is indeed difficult to restrain one's imagination and not to hazard some conjectures on the causes that could have led to such great effects. Besides, has not geology been so fertile in systems of this kind, for more than a century, that many people believe that they essentially constitute it, and regard it as a purely hypothetical science? What we have said hitherto shows that geology has a part that is as positive as any other observational science; but we believe we have shown at the same time that this positive part is as yet far from complete, that it has not yet collected enough facts to furnish an adequate basis for explanations. In the present state of the sciences, explanatory geology is still an indeterminate problem, for which no solution will be superior to others, until it has a larger number of fixed conditions. Meanwhile, systems have had the merit of giving an incentive for research into facts, and in this respect we ought to acknowledge their authors.

Those of Woodward, Whiston, Burnet, Leibniz, and Scheuchzer have long been known: conceived before we had any detailed notion of the structure of the globe, they could not sustain any serious examination. Buffon's first system [1749] eclipsed them all by the eloquent manner in which it was presented: it excited a general enthusiasm, and in a way it produced observers in every corner of the globe. One was thus really indebted to it, even for the observations that would destroy it. The second system by the same author, presented still more artfully in his "Epochs of nature" [1778], came too late to have even a momentary success. All naturalists were [by then] animated by the true spirit of observation, the search for positive facts; and one might say that since then those who have put forward ideas on these great subjects have been speculative geniuses or bold contemplatives, rather than philosophical observers.

34. [A secular version—with the usual personified Nature—of the view commonly expressed in Britain at this time, that the ancient coal deposits had been stored for eventual human use, by the care of divine providence.]

35. [The word is used here in its original—and by Cuvier's time archaic—sense, to mean anything distinctive dug up out of the ground, as in the still extant phrase "fossil fuels."]

The most indisputable consequences of the facts will already have yielded plenty to frighten minds habituated to the rigorous—or, if one will, timid—course that the sciences follow today. The primitive diminution of the waters, their repeated returns, the variations of the materials they deposited and which now form our strata; those of the organisms whose remains fill a part of these strata; the first origin of these same organisms: how are such problems to be resolved with the forces that we know now in nature? Our volcanic eruptions, our erosions, our currents, are pretty feeble agents for such grand effects: besides, there is nothing so violent that it has not been imagined. According to one [author], comets struck the earth, or consumed it, or covered it with the vapors in their tail. Others have supposed that the earth came out of the sun, either as liquid glass or as vapor. Chasms have been placed in its interior, which would have collapsed successively, or emanations have been made to escape violently from them. One has gone as far as to believe that its mass could have been formed by the union of fragments of other planets. Whatever the talent and force of mind that was needed to imagine these systems and to make them tally with the facts, we cannot include them in this display of the progress of the sciences: they tend rather to block true progress, by allowing one to believe that one can dispense with continuing observations on a matter that is so important and yet scarcely grazed.[36]

Translated from Cuvier, Rapport historique *(1810), pp. 131–51; the subheadings were printed in the margin. In the notes to this text no attempt has been made to identify the large number of authors who are merely mentioned in passing; likewise their publications are identified (and listed in the bibliography) only if Cuvier cited their titles specifically.*

36. The most complete historical account to have appeared in French, on the various systems imagined by geologists, is found in the *Theory of the earth [Théorie de la terre]* of Mr. de Lamétherie, Paris, [2nd ed.,] 1797, 5 vols., 8vo, a work that also contains the most methodical collection of the facts that geology comprised at the time that it was published. To it should be added those by Messrs. de Marschall, Bertrand, Lamarck, André de Gy, Faujas de Saint-Fond, and others that have appeared since that time.

THE GEOLOGY OF THE REGION

AROUND PARIS

———————

The preparation of his report on geology obliged Cuvier to become well informed—if he was not already—about recent and current research in that field. In particular, his review of work on the succession of formations in specific regions must have led him to appreciate what such a survey could do for his own research. He had entered the field of fossil anatomy with the scarcely examined assumption that *all* fossil bones dated from the *same* "catastrophe." As those from around Paris grew in importance, to become almost the centerpiece of his anatomical research, he became aware that they were decidedly more ancient than those of the mammoths and similar animals: the Paris fossils came from within a thick pile of bedded rocks, whereas the others were confined to loose superficial deposits. His tentative suggestion (text 6) that fossil bones became more unlike those of living animals as they were traced back into more ancient formations depended on some sense of their relative ages. But that sense could only come from attending to the work of those who had carefully plotted the relative positions of rock formations.

At some point—probably a year or two before his first lectures on geology (chapter 8)—Cuvier had begun collaborating with Alexandre Brongniart on a study of the rocks around Paris, in order to fix more clearly the geological context of the fossil bones that had now become a major focus of his research. Brongniart, Cuvier's almost exact contemporary, had been trained as a pharmacist, and he published some competent work in zoology. But during the Revolution he was appointed a mining engineer and taught at the École des Mines (Mining School) in Paris; so he had turned increasingly to chemistry and mineralogy. In 1800, at the age of only thirty, he had been appointed director of the state porcelain factory at Sèvres, just outside Paris; so he had his own reasons for wanting to explore the Paris region, in search of new sources of ceramic materials, particularly kaolin. He was familiar with German work on "Geognosie," and had realized that the formations around Paris were not included in the standard sequence of strata proposed by Werner. He and Cuvier therefore set themselves to work out the succession of those formations—in modern terms, their stratigraphy—in Wernerian manner; but it is not surprising that they paid much more attention to the fossil contents of the formations than was usual among the Germans.

Their working methods are difficult to reconstruct from the scanty evidence that survives, though it seems that both worked in the field, at times separately, at times together. Unlike any fieldwork that would have taken him farther from Paris, this project was compatible with the demands of Cuvier's various official duties; it was the only substantial piece of geological fieldwork he ever did. However, the field research was mainly due to Brongniart, as Cuvier in fact later acknowledged, with uncharacteristic generosity. The modest and self-effacing Brongniart allowed his collaborator's name to stand before his own, when their work was published; but the painstaking local details, the descriptions of minerals, and the comments on the practical uses of some of the rocks are probably all his. On the other hand, the manuscript describing their joint research is in Cuvier's hand; and he may have added the comments on the causal interpretation of each formation, which give the impression of having been tacked on to the more descriptive parts of the text.

A preliminary report on their research was presented to the Institut in 1808 (text 15), illustrated by a colored geological map of the region around Paris. The paper was published in the *Annales du Muséum* and in the *Journal des mines;* the latter, the organ of the Council of Mines, gave it a wide circulation in France and beyond (an English translation was published only two years later, despite the major war between the two countries). The revised version of the paper, amplified to book length by detailed local descriptions, was published in the *Mémoires* of the Institut in 1811, now accompanied by the geological map (see fig. 17) and by a set of geological sections drawn accurately to scale (see fig. 18); it was reissued in Cuvier's *Ossemens fossiles* the following year.[1]

The title of Cuvier's and Brongniart's paper placed it at once in the German tradition: "mineral geography" was a standard equivalent of "Geognosie." But even the first paragraph showed how it aimed to transcend that tradition: it placed fossils at the center of the stage, reporting an alternating sequence of marine and freshwater mollusks; and it proposed a causal explanation of the present topography, attributing it to the erosive effects of a major "irruption." Most of the paper, however, was primarily descriptive. It is translated here in full, not only because it is a major classic of geology that is cited far more often than it is read, but also to convey a sense of the empirical foundations on which Cuvier sought to ground his theoretical conjectures in geology.

After an introduction that describes the topography of the area, Cuvier and Brongniart describe nine formations, primarily in order from the lowest to the highest in the pile, or, in other words, in the true order of their relative age (fig. 16).

The lowest and oldest formation was the Chalk *(craie),*[2] a distinctive soft white limestone that was well known from many areas in northwest Europe (Cuvier himself had described it in Normandy: see text 1). Its equally distinctive fossils—of obviously marine organisms—indicated that it had been deposited on an ancient sea bottom. The Chalk formed the surface of the present land in a broad ring around Paris, and was known to extend at depth below the city; it was therefore evident that its upper surface was in the shape of a shallow basin. All the other formations had been deposited in succession within this basin, in what was taken to have been alternately a large lake and a gulf of the sea.[3]

1. Even the full version of the memoir, with the map attached, antedated William Smith's celebrated geological map of England and Wales (1815). The relation between the English and French projects was the subject of much chauvinistic argument among earlier historians of geology, anxious to claim national credit for the first modern-style geological map. Brongniart—but not Cuvier—had visited London in 1802 during the Peace of Amiens, and is likely to have seen a draft of Smith's map and heard of his fossil-based methods. Smith's map covered a much larger area than the French map.

2. The initial capital serves to distinguish the formation of the *Chalk* from the material *(chalk)* of which it is mainly composed; as Cuvier had long known (see text 1), it also contains lumps of flint, a totally different material. This modern convention will be adopted here for the sake of clarity; in text 15, on the other hand, the authors' convention is followed (e.g. "chalk" for "craie").

3. In contrast to the later nineteenth-century (and modern) interpretation of what is still known as the Paris Basin *(bassin de Paris),* Cuvier and Brongniart did not think the Tertiary (post-Chalk) formations had themselves been folded in any way. For them it was, quite literally, a *basin* of Chalk that had been filled, layer by layer, with a series of formations, deposited in alternately marine and freshwater conditions, until in some final event the area had emerged as dry land for the last time, and the present valleys had been scoured out.

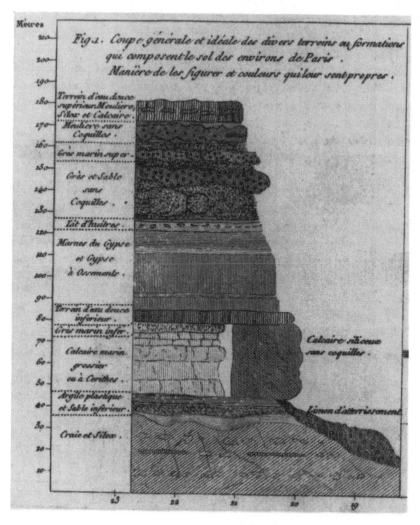

FIGURE 16 The "ideal section" with which, in their full report (1811), Cuvier and Brongniart summarized the Parisian formations in visual terms: "General and ideal section of the different *terrains* or formations that compose the ground *[sol]* of the region around Paris." The names of several formations embodied their interpretation of the sequence as a product of alternating marine *(marin)* and freshwater *(d'eau douce)* deposition. The Coarse Limestone and the Siliceous Limestone are placed strikingly side by side, as lateral equivalents. Cuvier's strangest mammals, the palaeotherium and anoplotherium, came from the gypsum formation in the middle of the section, whereas his fossil elephants and so forth (chapter 9) came from the far younger and "superficial" Detrital Silt *(limon d'atterrissement)*, lying in the valleys eroded in the pile of "regular" formations.

Resting on the Chalk, and in striking contrast to it, was the Plastic Clay *(argile plastique)*, widely used for making pottery and finer ceramics. Its very sharp and irregular junction with the Chalk was interpreted as marking an abrupt change of conditions and perhaps also a great lapse of unrecorded time. Although the authors could not report finding any fossils whatever in the Plastic Clay, they implied that it had been deposited in freshwater.

Next came the Coarse Limestone *(calcaire grossier)*, which was extensively quarried as a building stone for Paris. Cuvier and Brongniart claimed to be able to recognize some of its subdivisions over long distances, by the "characteristic fossils" found in specific beds.[4] More importantly, its extremely diverse and well-preserved fossils—said to be of some six hundred species—were quite different from those of the Chalk, and most were also different from any species known alive (the authors here relied on the careful taxonomic work of Lamarck, who was currently describing and naming them in the *Annales du Muséum*). Since the fossils were unquestionably of marine genera (see fig. 19), the Coarse Limestone was taken to mark a return to marine conditions (assuming the Plastic Clay had in fact accumulated in freshwater).

Such an alternation was made more plausible in the light of the Gypseous Formation *(formation gypseuse)* that overlay the Coarse Limestone. This was the formation quarried for building stone and for its gypsum, in which many of Cuvier's more puzzling vertebrate fossils were found (see chapters 4, 6, 7). Shells of freshwater mollusks, although rare, indicated that it had been deposited in a lake or in marshy conditions. Immediately overlying the gypsum, however, was a thin but widespread bed of clay with marine shells, which the authors took to mark an abrupt reversion to marine conditions (see figs. 18, 19).

That interpretation was confirmed by the overlying formation of Marine Sandstone *(grès marin)*, for it contained fossil shells of the same species as those of the Coarse Limestone. So the authors' emerging interpretation was summarized at this point in terms of three periods of marine conditions alternating with two of freshwater. The changes were assumed—without any special reasoning—to have been sudden.

The formation of Siliceous Limestone *(calcaire silicieux)* described next was in fact out of order, and in a position that the authors evidently found puzzling. For it occupied the *same* position in the sequence as the

---

4. Unlike William Smith, they used "characteristic fossils" not to distinguish the different formations—at least not explicitly—but only on a much more restricted scale, for recognizing specific subdivisions within this one formation. In contrast to Smith, they used fossils primarily for what would now be called *paleoecological* purposes (see below).

Coarse Limestone, in parts of the Paris region where that formation was missing. That two distinct formations should thus be (in modern terms) lateral equivalents of each other was quite unexpected in geognostic practice; and Cuvier's and Brongniart's unease about the situation is reflected in the awkward way they represented it in the "ideal section" that later summarized the whole pile of formations (see fig. 16).

The next formation, of Sandstone without Shells *(grès sans coquilles)*, overlay all those described previously. It covered wide areas, notably in the forests near Fontainebleau, southeast of Paris; it was worked for its fine sands, but in the absence of fossils nothing was said about its conditions of deposition.

The last of the "regular" formations to be described was given a name that embodied just such an interpretation: it was the Freshwater Formation *(terrain d'eau douce)*. It contained shells like those now living in freshwater, but the authors claimed that no similar sediments were accumulating any longer: although unequivocally of freshwater origin, it still belonged to "the ancient world." That inference seemed to be confirmed by its position: it capped what were now some of the highest plateaus around Paris, and therefore predated the excavation of the valleys.

Conversely the ninth and last formation, the Detrital Silt *(limon d'atterrissement)*, was loose, superficial, and confined to the floors of river valleys; in it were found the bones of Cuvier's fossil elephants and other such mammals. Clearly it dated from *after* whatever event or process had formed the present topography of the Paris region: the older formations had been deeply eroded into the valleys in which the present rivers flow, before this material was deposited (see fig. 18). So although it was termed "very modern" in comparison to all the other formations, it too was placed firmly in "the ancient world," and the authors claimed that it was separated from historic times by a "long succession of centuries."

This impressive paper applied the well-established fieldwork methods of geognosy to a succession of formations more recent than any that had hitherto been described, at least so thoroughly. Even the Chalk, the *oldest* formation considered, had generally been regarded as being among the youngest known. Cuvier and Brongniart used the evidence of invertebrate fossils to reconstruct a complex history of alternating marine and freshwater conditions, at least for the Paris region. But above all, the paper provided a context of earth history in which Cuvier could place his fossil vertebrates. It showed that the best known of them—such as the mammoths—were extremely recent in geological terms, though still implicitly belonging to a prehuman world. Those he had worked on from around Paris, which were much less similar to any living mammals, were demonstrably far older. Here was a starting point, at least, for constructing firmer foundations for Cuvier's earlier bold claim about the succession of vertebrate life.

TEXT 15

⎯

Essay on the Mineral Geography of the Environs of Paris

By Mr. Cuvier and Mr. Alexandre Brongniart

[INTRODUCTION]

The region in which this capital is situated is perhaps one of the most remarkable that has yet been observed, in the succession of varied formations *[terrains]* that it comprises, and from the extraordinary remains of ancient organisms that it conceals. Thousands of marine shells, alternating regularly with freshwater shells, make up the principal mass of it. The bones of terrestrial animals, of which even the genus is entirely unknown [alive], fill certain parts of it; other bones of species that are notable for their size, and of which we find only a few congeneric species in very distant lands, are scattered in the most superficial beds. A strongly marked trace of a great irruption, coming from the southeast, is imprinted on the form of the bluffs and the direction of the valleys. In a word, there is no region more capable of instructing us on the latest revolutions that have ended the formation of our continents.

However, this country has been very little studied from this point of view. Although it has been inhabited for so long by so many educated men, what has been written on it boils down to some fragmentary essays. Almost all of them are either purely mineralogical, without any regard for the organic fossils, or purely zoological, and without regard for the position of the fossils. A report by Lamanon on the gypsums and their bones is perhaps the only exception to this classification. We should, however, acknowledge that the excellent description of Montmartre by Mr. Desmarest; the information on the basin of the Seine, given by the same savant in the *Methodical encyclopedia;* the mineralogical essay on the Department of Paris, by Mr. Gillet-Laumont; the extensive and beautiful research on the fossil shells of its environs, by Mr. de Lamarck; and the geological description of the same region, by Mr. Coupé, have [all] been consulted by us with profit, and have often guided us in our travels.[5] Nevertheless, we think that the work of which we here present a first sketch will not be without interest, [even] after all those we have just cited.

Having started four years ago, and having continued with much effort, making numerous journeys and collecting information and specimens everywhere, we are far from thinking it is yet completed; and above all we ask that the summary of it that we are about to read should not be confused with the detailed version that we shall publish soon. Certain circumstances oblige us to present this summary today, and to reserve a time for some equally long and laborious researches, before the happy moment when we will believe we have brought them to completion.[6]

[OUTLINE OF THE PARIS BASIN]

By the nature of their object, our routes have had to be limited according to the nature of the terrain, and not by arbitrary distances. Thus we first had to determine the physical limits of the region that we wanted to study.

The basin of the Seine is separated over a fairly wide space from that of the Loire, by a broad elevated plain, the largest part of which has the common name of Beauce; the middle and driest portion of this extends from northwest to southeast, over a space of more than forty leagues [120 miles], from Courville to Montargis. This plain slopes to the northwest to a higher and above all more hilly country, where the rivers Eure, Avre, Iton, Risle,

---

5. [Robert de Paul de Lamanon (1752–87), had been an active Parisian naturalist until his tragically early death while on la Peyrouse's famous expedition in the Pacific; in his paper "Fossiles trouvés dans les carrières de Montmartre" (Fossils found in the Montmartre quarries, 1782) he had suggested a former lake on the site of Paris during the deposition of the gypsum. Nicolas Desmarest (1725–1815) had been prominent in mineral natural history ever since the publication of the great *Encyclopédie*, and was famous for his classic descriptions of the extinct volcanos of Auvergne: his paper "Couches de la colline de Montmartre" (Beds of the hill of Montmartre, 1804) had drawn a parallel between the prismatic jointing of the gypsum and that of many basalts; and he had already published three volumes on "géographie physique" (1794–1806)—which included much that had come to be called "geology"—in the *Encyclopédie méthodique*. Lamarck had almost completed his long series of papers "Fossiles des environs de Paris" (1802–9), describing the mollusks. Jacques Michel Coupé (1737–1809) had been active in revolutionary politics before turning to geology; his paper "L'étude du sol des environs de Paris" (A study of the ground in the Paris region, 1805) was one of very few previous attempts to describe the pile of Parisian formations.]

6. [It is possible that the unidentified "circumstances" that precipitated this preliminary account of Cuvier and Brongniart's work were rumors that Smith's map of England and Wales was about to be published; John Farey, one of Smith's English colleagues, later protested in print that the Frenchmen were trying to steal the credit from Smith. But a more likely explanation is that Brongniart, who was a candidate for the professorship of mineralogy at the Collège de France, needed to have his research in the public realm as soon as possible, before the decision was made (he was in fact appointed). This paragraph, which forms an integral part of the original manuscript, was included in the text published in the *Annales du Muséum* but omitted from the *Journal de physique*.]

Orne, Mayenne, Sarthe, Huisne, and Loir have their sources. The highest part of this country, between Sées and Mortagne, once formed the province of the Perche and a part of Lower Normandy, and belongs today to the department of the Orne. The physical line of separation between the Beauce and the Perche passes roughly by the towns of Bonneville, Alluye, Illiers, Courville, Pontgouin, and Verneuil. On all the other sides, the plain of Beauce dominates what surrounds it. Its descent on the side of the Loire does not concern us; that on the side of the Seine is composed of two lines, one to the west which looks toward the Eure, the other facing toward the Seine. The first goes from Dreux toward Mantes; the other starts from near Mantes, passes by Marly, Meudon, Palaiseau, Marcoussy, La Ferté-Alais, Fontainebleau, Nemours, etc. But these two lines should not be represented as straight or uniform; on the contrary they are endlessly uneven and broken up, in such a way that if this vast plain were surrounded by water its coasts would show gulfs, capes, and straits, and would be surrounded everywhere by islands and islets.[7]

Thus in our vicinity the long plateau on which are the forests of Saint-Cloud, Ville-d'Avray, Marly, and Les Aluets, and which extends from Saint-Cloud up to the confluence of the river Maulde with the Seine, would be an island separated from the rest by a channel on the present site of Versailles, the little valley of Sèvres, and the large valley of the park at Versailles. The other upland, in the form of a fig leaf, on which are Bellevue, Meudon, and the forests of Verrière and Chaville, would form a second island separated from the mainland by the valley of Bièvre and that of the hills of Jouy. But then, from Saint-Cyr to Orléans, there is no further complete interruption, although the valleys of the rivers of Bièvre, Ivette, Orge, Étampes, Essonne, and Loing cut deeply into the mainland from the east side, and those of Vesgre, Voise, and Eure from the west. The most jagged part of the coast, which would have the most reefs and islets, is that commonly known as "Gâtinois français," and above all the part that composes the Forest of Fontainebleau.

The slopes of this immense plateau are generally quite steep. All the escarpments one sees there, as well as those of the valleys, and the wells sunk in the uplands, show that its physical nature is the same throughout. It is formed of a prodigious mass of fine sand, which covers this whole surface, passing on to all the other formations or lower plateaus that this large plain

7. [At a time when topographic maps indicated relief by crude hachuring, not contour lines, this imagined effect of submersion was an effective way of describing the topography of the Paris region, particularly since most readers would have been familiar with the towns, rivers, and forests mentioned by name.]

dominates. Its margin, which looks onto the Seine from La Mauldre to Nemours, will thus form the natural limit of the basin that we have to examine.

Immediately from beneath its two extremities, that is to say, toward La Mauldre and a little beyond Nemours, emerge two portions of a plateau of chalk that extends in all directions and to a great distance, to form the whole of Upper Normandy, Picardy, and Champagne. The internal edges of this great belt, which pass on the east side by Montereau, Sézanne, and Épernay, and on the west by Montfort, Mantes, Gisors, Chaumont, to near Compiègne, making to the northeast a considerable angle that covers the whole of the Laon area, complete—with the sandy flank just described—the natural limits of our basin.

But there is this great difference: the sandy plateau that comes from the Beauce is higher than all the others, and consequently more modern, and it ends completely along the edge we have marked; whereas on the contrary the chalk plateau is naturally more ancient and lower than all the others. It only ceases to appear beyond [i.e. inside] the line of the circuit we have just indicated; but far from finishing there, it visibly sinks under all the other [formations]; it is found again wherever the latter are penetrated deeply enough, and it even rises up again in some areas and, as it were, reemerges in piercing through them.[8]

One can thus say that the materials composing the Paris Basin, in the sense that we define it, have been deposited in a vast hollow space, in a kind of broad gulf of which the coasts were of chalk. This gulf perhaps made a complete circle, a kind of large lake; but we cannot be certain of this, given that its border to the southeast, and the materials it contained, has been hidden by the great sandy plateau we spoke about above. Besides, this great sandy plateau is not the only one to cover the chalk. There are several in Champagne and Picardy which, although smaller, are of the same nature and could have been formed at the same time. Like it, they rest directly on the chalk, in areas where the chalk was high enough not to be covered at all by the materials of the Paris Basin.

We shall first describe the *chalk*, the oldest of the materials we have in our vicinity. We shall end with the sandy plateau, the newest of our geological products. Between those two extremes we shall deal with the less extensive but more varied materials that filled the great cavity of the chalk, before the sandy plateau was deposited on the former as on the latter.

8. [The authors clearly understood the three-dimensional structure quite well; if their description seems confusing, it is because they lacked the technical vocabulary of slightly later (and modern) geologists, such as "outlier," "inlier," etc.]

These materials can be divided into two tiers [*étages*].[9] The first, which covers the chalk everywhere it was not elevated enough, and which has filled the entire floor of the gulf, is itself subdivided into two parts equal in level, placed not one above the other but end to end, namely: the plateau of nonshelly siliceous limestone, and the plateau of coarse shelly limestone. We know the limits of this tier fairly well on the side of the chalk, because the chalk does not cover it at all; but these same limits are masked in several places by the second tier, and by the great sandy plateau that forms the third [tier] and that covers a great part of the other two.

The second tier will be termed *gypseo-marly* [*gypso-marneux*]. It is not spread generally, but only from place to place or in patches; these patches are again very different from one another in thickness and in the details of their composition.

These two intermediate tiers as well as the two extreme tiers are covered in turn, and all the gaps they have left are in part filled, by a fifth kind of formation [*terrain*], mixed also with marl and chert [*silice*], which we term *freshwater formation* [*terrain d'eau douce*], because it teems exclusively with freshwater shells.

We have the honor to present the Class with a first attempt at mineralogical maps in which each kind of formation is highlighted by a particular color: the sand in fawn, the gypsum in blue, the shelly limestone in yellow, the siliceous limestone in purple, the chalk in pink, the freshwater formation in green striped with white. The rolled sands or alluvium, which have not been deposited at all quietly, but brought from elsewhere by the rivers, are marked in plain green; and in dark brown the peaty formations along the streams and around the ponds. This map, which is one of the main results of our travels, is perfectly accurate where it is colored, and we have left blank what we do not yet know sufficiently (fig. 17).[10]

Such are the large masses of which our region is composed, and which form its different tiers. But by subdividing each tier one can reach still greater precision, and obtain more rigorous mineralogical results. This gives up to six distinct kinds of beds [*couches*], which we are first going to enumerate briefly.

9. [The word "étage" has structural or spatial connotations, being used for example for the floors or stories of a building. As such it was a highly appropriate term to use in this kind of geognostic description; it does not have the geohistorical meaning of the word "stage" in its modern geological usage.]

10. [This paragraph of the manuscript text was published in the *Annales du Muséum* but omitted from the version in the *Journal des mines*. It shows that Cuvier and Brongniart's map, or at least a preliminary version of it, was in a sufficiently advanced state to be exhibited at the Institut in 1808, three years before it was published. The authors' claim that theirs was the "first attempt" at a map of this kind was what later angered Smith's supporters. (The coloring of the copy reproduced here does not match the colors described in the text.)]

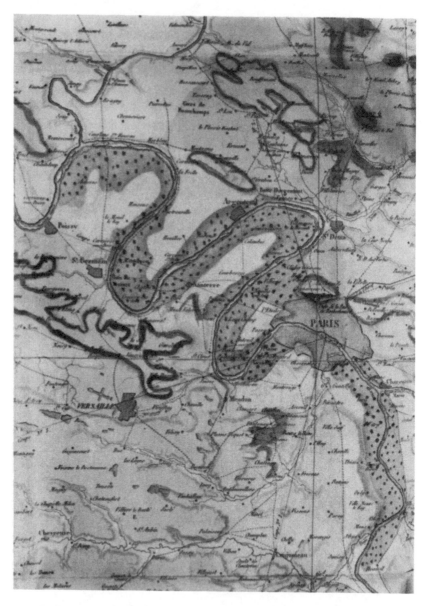

FIGURE 17 A small part of the "geognostic map" of the Paris region published by Cuvier and Brongniart in 1811, but displayed at the Institut—perhaps in a preliminary form—as early as 1808. The hand-colored original distinguishes twelve different formations: in this reproduction the dark loops indicate outcrops of the Gypseous Formation on the flanks of hills capped with the overlying Marine Sandstone and Sandstone without Shells; the speckled areas denote the superficial Detrital Silt in the valley of the Seine. The straight lines radiating from the center of Paris mark the positions of the accompanying sections (see fig. 18).

1. CHALK FORMATION [CRAIE]

In the Paris region, as in almost all the places it has been observed, the chalk forms a mass in which the layers [assises] are often so indistinct that one would almost doubt that it was formed in beds [lits], if one did not see these discontinuous seams of flint which, by their perfectly horizontal position, parallelism, continuity, and frequency, indicate successive and almost periodic deposits. Their distance apart varies with the locality: at Meudon they are about two meters from each other, and the space between any two beds of flint contains no isolated piece of that stone. At Bougival the seams are distant and the flints much less numerous. The chalk that contains them is not pure carbonate of lime. According to Mr. Bouillon-la-Grange, it contains about 11% of magnesium and 19% of silica, the greatest part of which is in the form of a sand that can be separated by washing.

The fossils found in it are not numerous, compared to those that are seen in the beds of coarse limestone that almost directly overlies the chalk; but they are entirely different from those fossils, not only in species but even in genera. Combining those we ourselves have observed with those collected by Mr. Defrance, we bring to fifty the number of species of fossils that we know from the chalk of the areas [terrains] that are the object of our study.[11]

The species of these fossils have not yet all been identified. We will list them and give exact identifications in our detailed monograph. Here we will content ourselves with saying that there are: two lituolites;[12] three tube worms; some belemnites that, according to Mr. Defrance, are different from the one that accompanies the ammonites in the compact limestone;[13] some shell fragments that, by their tabular form and fibrous structure, can belong only to the genus *Pinna*;[14] one mussel; two oysters; one species of the scallop genus; one crania; three terebratulas; a spirorbis; some echinoids [ananchites] in which the shelly envelope has remained calcareous and has taken on a crystalline texture whereas the interior alone has changed into flint; some porpytes; five or six different corals, one of which seems to be-

11. [Jacques Louis Marin Defrance (1758–1850) was a wealthy Parisian naturalist with an outstanding fossil collection.]

12. [Small chambered shells, at this time thought to be related to those of cephalopod mollusks such as the pearly nautilus, but later (and by modern paleontologists) attributed to foraminifera or shelled protozoans. Most of the other names in the following list have not changed their meanings substantially and need no special explanation; all are Chalk fossils that are familiar to modern paleontologists.]

13. [The reference is to one or more of the harder limestones (in modern terms, mostly of Jurassic age) that were known to underlie the Chalk and to outcrop farther from Paris.]

14. But if one infers the size of the individuals to which they belonged from the thickness of these fragments, one concludes that these shellfish were monstrous: we have measured pieces that are twelve millimeters thick, although the thickness of the largest-known [living] species of *Pinna* is only

long to the genus *Caryophyllaea* and another to the genus *Millepora*, the latter usually being brown and in the state of iron oxide as a result of the decomposition of pyrite; and finally some sharks' teeth.

We should note, with Mr. Defrance, that no univalve shell with a simple and regular spiral has yet been found in the chalk. This fact is all the more remarkable, since we shall meet with these shells in great abundance, some meters above the chalk, in beds that are likewise calcareous but with a different structure.

Among the quarries and hills of chalk that we have visited, we will cite *Meudon*. The chalk there is not bare, but is overlain by the plastic clay and the coarse limestone. The upper part of this mass is as if broken, and displays a kind of breccia in which the fragments are of chalk and the matrix of clay.

The highest part of the mass of chalk seemed to us to be above the Sèvres glassworks. It is fifteen meters above the Seine. This position raises all the formations *[couches de terrain]* that overlie it, and at the same time seems to diminish their thickness. The mass of stone dips noticeably toward the river.

At *Bougival*, near Marly, the chalk is almost bare at some points, being covered only by some fairly fine-grained calcareous rocks, but in more or less coarse fragments scattered in a marly sand that is almost pure toward the summit. In the middle of these fragments there are geodes of a pale yellowish, compact, fine-grained limestone, with crystalline layers and little cavities lined with tiny crystals of calcite *[chaux carbonatée]*. The matrix of these geodes contains an abundance of little spiral univalve shells, which seems to prove that this limestone does not belong to the formation of the chalk.

Among these geodes we found one that has a wide cavity lined with clear, elongated, and sharp crystals, more than two centimeters long. Mechanical cleavage alone taught us that these crystals belonged to the species of strontium sulfate, and a closer examination of their forms showed us that they constituted a new variety.[15] Mr. Haüy, whom we have informed of this, has named it *apotome* strontium sulfate. These crystals show four-faced rhomboidal prisms, the angles of which are the same as those of the prisms of the *unitaire, émoussée,* etc. varieties, namely $77° 2'$ and $102° 58'$. They are terminated by very acute four-faced pyramids. The angle of incidence

---

two millimeters. [This note was printed as a long parenthesis, not as a footnote. Here and elsewhere in this paper, the metric system of measurement—devised and adopted during the Revolution—is adopted in place of traditional units such as fathoms *(toises)*, leagues *(lieues)*, etc. (see also figs. 16, 19).]

15. [Like other naturalists at this time, Cuvier and Brongniart use the terms "species" and "variety" *(espèce, variété)* for the classification of both minerals and organisms: in a nonevolutionary

of the faces of each pyramid onto the adjacent [prism] face is 161° 16'. The faces are produced by virtue of a decrease by twos, arranged to the left and to the right of angle E of the primitive form. This is a law that had not yet been recognized in the varieties of strontium sulfate studied until now. Its sign will be $^1E\ E^2\ ^2E$.[16] The crystals of strontium sulfate observed hitherto in the Paris region are extremely small, and line the interiors of some of the strontian geodes that are found in the green shales of the gypsum formation; but they had never before been seen so clearly and in such volume.

2. PLASTIC CLAY FORMATION [ARGILE PLASTIQUE]

Almost all the surface of the mass of chalk is covered with a bed of plastic clay of remarkably constant character, though at various points it shows detectible differences. The clay is greasy and tenacious; it contains silica but very little lime, such that it gives no effervescence with acids. It is even absolutely infusible in a porcelain furnace as long as it does not contain too much iron. It varies much in color: it can be pure white (at Moret, in the Forest of Dreux), gray (at Montereau, Houdan, and Condé), yellow (at Houdan and at Abondant in the Forest of Dreux); pure slate gray, slate gray mixed with red, or almost pure red (all south of Paris, from Gentilly to Meudon).

This plastic clay is used, according to its various qualities, to make either china [faïence fine] or stoneware [grès], or porcelain crucibles and boxes, or even earthenware [poterie rouge] that has the hardness of stoneware when it can be fired appropriately. It is never either effervescent or fusible. Its only defects are the red color, grains of pyrite, bits of flint, little fragments of chalk, and crystals of selenite that are sometimes found in it.

This bed [couche] varies greatly in thickness: in some parts it is up to sixteen meters or more; in others it forms only a thin layer [lit] one or two decimeters [10–20 cm] thick.

It seems almost certain that no fossil, either marine or terrestrial, is found in this clay; at least we have seen none, either in the different beds that we have observed in place, or in the considerable masses that we have examined on several occasions in the numerous manufactories that make use of it; and finally the workmen who exploit the clay to the south of Paris have assured us that they have never met with either shells or bones or wood or plants there. Dolomieu, who has recognized this same formation

---

framework, the same terms could be applied equally well to natural kinds in any of the three "kingdoms" of nature.]

16. [This description adopts the rigorous quantitative methods and classificatory notation of Haüy's crystallography. Most of his "varieties" were distinctive crystal forms, with exactly the same angles between the crystal faces.]

[banc] of clay between the chalk and the coarse limestone in the loop that the Seine makes opposite Rolleboise,[17] says correctly that fragments of bituminous wood were found there and had even been mistaken for coal; but he observes that these little pieces of lignite were found in collapsed parts of the bed, which could have enveloped them at an epoch later that the original deposition of this clay.

The localities we have cited above prove that this bed of clay is of very wide extent, and that it conserves the principal characters of its formation and position throughout that extent.

If we compare the descriptions we have just given of the beds of chalk and those of plastic clay, we notice (1) that not only are none of the fossils met with in the chalk found in the clay, but that no fossils at all are found in it; and (2) that there is no gradual passage at all between the chalk and the clay, since the parts of the bed of clay closest to the chalk contain no more lime than the other parts.

It seems to us that one can conclude from these observations, first, that the liquid that deposited the bed of plastic clay was very different from that which deposited the chalk, since it contained no perceptible carbonate of lime, and that in it lived none of the animals that inhabited the waters that deposited the chalk. Second, that there was necessarily a distinct separation, and perhaps a long span of time, between the deposition of the chalk and that of the clay, since there is no transition between the two kinds of formation. The kind of breccia that we have noticed at Meudon, with fragments of chalk in a matrix of clay, even seems to prove that the chalk was already solid when the clay was deposited. This material *[terre]* penetrated between the fragments of chalk produced at the surface of the chalky ground by aqueous movement or by quite another cause.

The two kinds of formation we have just described have thus been produced in utterly different and even very sharply contrasted circumstances. They are the result of the most distinct and most characteristic formations *[formations]* that can be found in geognosy, since they differ in chemical nature, in the kind of stratification, and above all in that of the fossils that are found in them.

3. SAND AND COARSE LIMESTONE FORMATION *[CALCAIRE GROSSIER]*

The coarse limestone does not always overlie the [plastic] clay directly; often it is separated by a more or less thick bed of sand. We cannot say whether this sand belongs to the limestone formation or to that of the clay.

---

17. *Journal of mines*, no. 9, p. 45 [Dolomieu, "Observations sur la prétendue mine de charbon de terre" (1795)].

In the few places where we have observed it, we have not found any shells that would connect it with the clay; but since the lowest limestone bed is ordinarily sandy and always full of shells, we do not yet know if this sand is different from the former or if it is the same deposit. What would make us suspect that it is different, is that the sand of the clays that we have seen is generally fairly pure, although red or bluish gray in color. It is refractory and often very coarse-grained.

The limestone formation following on from this sand is composed of alternating beds, of more or less hard coarse limestone, of clayey marl and even of clay foliated in very thin beds, and of limy marl. But it should not be thought that these various beds are placed there at random or without rules: they always follow the same order of superposition over the considerable extent of terrain that we have covered. There are sometimes several that are missing or are very thin; but the one that is lower in one area does not become the upper in another.

This constancy in the order of superposition of the thinnest beds, over an extent of at least twelve myriameters [120 km] is, in our opinion, one of the most remarkable facts that we have noted in the pursuit of our researches. The consequences that should flow from it, for the [practical] arts and for geology, should be as much more interesting as they are more certain.

The means we have used to recognize a bed *[lit]* already observed, in a far distant area, amid such a large number of limestone strata, is drawn from the nature of the fossils enclosed in each bed. These fossils are always generally the same in the corresponding beds, and show quite notable differences of species from one set *[système]* of beds to another set. This is a sign of recognition that so far has not misled us.

It should not be thought, however, that the difference from one bed to another is as sharp as that between the chalk and the limestone. If it were, one would have the same number of distinct formations *[formations particulières]*. But the fossils characteristic of one bed become less numerous in the higher one, and disappear altogether in the others, where they are replaced little by little by new fossils that had not appeared at all before.

Following this course, we shall indicate the principal sets of beds that can be observed in the coarse limestone. In our subsequent monograph a complete description will be found, stratum by stratum, of the many quarries that we have studied in order to reach the results that we present here in a general manner.

The lowest beds of the limestone formation are the best characterized. They are very sandy, and often even more sandy than calcareous. When they are solid, they decompose in air and disintegrate into powder; thus this stone cannot be utilized at all. The shelly limestone of which it is composed, and even the sand that sometimes replaces the stone, almost always encloses green earth in powder or in grains. From the analyses we have made, this earth is analogous in composition to the *baldogée* chlorite or Verona earth.[18] It owes its color to iron, and is found only in the lower beds. It is not seen in the chalk or in the clay, nor in the middle and upper calcareous beds; and its presence can be regarded as a sure indicator of the proximity of the plastic clay, and consequently of the chalk. But what characterizes this set of beds still more particularly is the prodigious quantity of fossil shells that it encloses. To give an idea of the number of species that these beds contain, it is sufficient to say that Mr. Defrance has found more than six hundred there, all of which have been described by Mr. de Lamarck.[19]

We should note that the majority of these shells differ much more from species now living, than those of the upper beds. Among the fossils particular to these lower beds, we shall mention pectuncles, solens, oysters, mussels, pinnas, calyptreas, pyrulas, large tellines with elongated edges, terebellas, porpytes, madrepores, and notably nummulites and fungites.[20] Such are the shells most characteristic of this bed. We ought to note that it is not [just] from the single deposit at Grignon that we have collected the examples we have just cited; those examples [alone] would hardly have characterized the set of beds that we want to make known. We chose them from the quarries at Sèvres, Meudon, Issy, Vaugirard, and Gentilly; and in the beds at Guespelle, at Lallery near Chaumont, etc.[21]

It is in this same bed that camerines are found, either on their own or mixed with madrepores and the shells mentioned above. They are always the lowest, and consequently the first that were deposited on the chalk formation; but they are not present everywhere. We have found them near Villers-Cotterêts, in the valley of Vauciennes; at Chantilly on the descent from the hill, where they are mixed with very well preserved shells and with coarse grains of quartz that make this stone a sort of conglomerate; at Mont Ouin near Gisors, etc.

18. [A green pigment used in painting, classed by Haüy as a variety of talc.]

19. [Lamarck, "Fossiles des environs de Paris" (1802–9).]

20. [A varied assemblage of mollusks, corals, etc. The more familiar were listed with their vernacular names; the less familiar, with French versions (here anglicized) of their scientific Latin names. Nummulites (the subject of one of Cuvier's few papers on invertebrate fossils) were thought to be small chambered mollusk shells, but were later interpreted as giant foraminiferan protozoans. The small "fungite" coral was one of the few fossils significant enough to be illustrated in the only plate in the full report to be devoted to fossil specimens (see fig. 19, lower right).]

21. [The authors here emphasize that their fossil localities are *not* those exploited by Lamarck (or at least not his most prolific quarry at Grignon): presumably this is to assert the independence of their research, to show the wide distribution of this subdivision of the Coarse Limestone, and to stress that no such set of beds can be characterized adequately from any single locality.]

Another notable character of the shells in this bed is that most of them are complete and well preserved, that they can easily be detached from the rock, and finally that many of them preserve their nacreous luster. In all the places mentioned above, and in other less remarkable ones, we have noticed that the sandy limestone beds that contain these shells follow immediately on the plastic clay that overlies the chalk. It is by these multiple observations that we have verified the generality of the rule we have just established.

The other sets of beds are less distinct, and we have not yet been able to determine the analysis of the numerous observations we have made, to establish precisely the succession of different fossils that characterize them. We can, however, report that after inspecting the quarries to the south and west of Paris, from Gentilly to Villepreux and Saint-Germain, the higher beds succeed those we have described in the following order.

1. A soft bed, often with a greenish tint, which the workers therefore call *green bed.* In its lower part it frequently shows the brown imprints of leaves and plant stems.

2. Gray or yellowish beds, sometimes soft, sometimes very hard, containing principally rounded venuses, ampullarias and above all tuberculate cerithia which are often present in prodigious quantity. The upper and middle part of this bed, often very hard, is used as a very good building stone, known under the name of *rock [roche].*

3. Finally and toward the top, a thin but hard bed, which is remarkable for the prodigious quantity of little elongated and striated tellinas. They occur on its horizontal bedding planes, lie flat and tight against each other, and are generally white.

Above these last beds of coarse limestone come hard calcareous marls; they split into fragments, the surfaces of which are normally covered with a yellow coating and with black dendrites. These shales are separated by soft calcareous shales, by argillaceous shales, and by calcareous sand that is sometimes cemented and contains flints in horizontal zones. We assign to this set the bed at the Neuilly quarries, in which quartz crystals and rhomboidal crystals of *inverse* carbonate of lime are found. But what is most characteristic of this last set of beds in the calcareous formation is the total absence of shells or any other fossils.

It follows from the observations we have just reported (1) that the fossils of the coarse limestone have been deposited slowly and in a calm sea, since they were deposited in regular and distinct beds; that they have not been mixed up at all, and that the majority are in a perfect state of preservation, however delicate their structure, so that even the tips of spiny shells are very often complete; (2) that these fossils are entirely different from those of the chalk; (3) that the longer the beds of this formation were deposited, the more the number of species of shells diminished, to the point at which they are no longer found. The waters that formed these beds either no longer enclosed them, or else lost the property of conserving them.

Certainly things happened in these seas quite otherwise than in our present seas: in the latter, beds are no longer forming, and the species of shells are always the same in the same areas. Since the time that oysters have been fished for on the coast of Cancale, for example, these shellfish have not been seen to disappear and be replaced by other species.[22]

4. GYPSEOUS [GYPSEUSE] FORMATION

The terrain of which we are now going to trace the [natural] history is one of the clearest examples of what should be understood by "formation" [formation]. We shall see in it beds that are very different from one another in chemical nature, yet clearly formed together.[23]

The formation that we term gypseous is not composed solely of gypsum; it consists of alternating beds of gypsum and argillaceous and calcareous marl. These beds follow an order of superposition that was always the same in the great gypseous belt we studied, which extends from Meaux to Triel and Grisy. Some beds are missing in certain areas, but the others always remain in the same respective position.

The gypsum lies immediately above the limestone, and this superposition is beyond doubt. The position of the gypsum quarries at Clamart, Meudon, and Ville-d'Avray, above the coarse limestone that is exploited in the same places; that of the quarries on the hill of Triel, where the superposition is still clearer; and finally a well sunk in the garden of Mr. Lopès at Fontenay-aux-Roses, which traversed first the gypsum and then the limestone: these are more than sufficient proof of the position of the gypsum above the limestone.

The gypseous hills and buttes have a distinctive aspect that can be recognized from a distance. Since they are always above the limestone, they

22. [Cuvier and Brongniart were not unusual, at this time, in insisting so forcefully on the contrast between past and present marine environments, although in other contexts they would probably have conceded—as others certainly did—that their knowledge of almost everything beneath low-tide level was abysmally scanty. The comment about the stability of the living oysters through human history, contrasted with the change observed in the fossil fauna, is perhaps an indication of the relatively short period of time the authors envisaged for the deposition of the Coarse Limestone. Cancale is on the north French coast near Saint-Malo.]

23. [The reference is to the "geognostic" definition of a "formation," as it had developed by the time that Cuvier and Brongniart were writing: namely, a collection of beds or strata that were evidently formed at the same period and under the same circumstances, even if the constituent rock types were quite diverse. As the French words noted in these translated texts make clear, the terminology of strata, beds, layers, formations, etc. was extremely fluid and imprecise at this time.]

are formed on the highest hills, as a second hill, elongated or conical, but always distinct.

We shall set out the details of this formation by taking as examples the hills that display the most complete ensemble of beds; and although Montmartre has already been much inspected, it is still the best and most interesting example we could choose [fig. 18].

At Montmartre, as in the hills that seem to follow on from it, three masses of gypsum can be recognized. The lowest mass is composed of thin alternating beds of gypsum (often selenitic), solid calcareous marls, and finely laminated argillaceous marls. It is in the first that coarse crystals of greenish lenticular gypsum are mainly seen, and in the last that opal is found. We know of no fossils in this [lowest] mass, which is the quarrymen's third [mass].[24]

The second or intermediate mass differs from the preceding one only in that the gypseous beds [bancs] are thicker, and the marly beds less frequent. Among these marls, one that should be noticed is argillaceous, compact, and marbled gray, and is used as a cleansing material [pierre à détacher]. It is mainly in this mass that fossil fish have been found. No other fossils whatever are known in it; but one begins to find strontium sulfate in scattered nodules [rognons, i.e. "kidneys"] in the lower part of the marbled marl.

The upper mass, which the workmen call the first, is in every respect the most remarkable and most important. It is also much more prominent than the others, for in some areas it is up to twenty-five meters thick. It is modified by only a small number of marly beds; and in some places, as at Dammartin and Montmorency, it is situated immediately below the soil. The lowest beds of gypsum in this first mass contain flints that seem to merge into the gypseous matter and to be penetrated by it. The middle beds split naturally into bulky prisms with several faces. Mr. Desmarest has described and illustrated them very well: they are called "tall pillars" [hauts piliers].[25] Finally, the highest beds are penetrated with marl; they are not very prominent, and alternate with beds of marl. There are normally five, which extend to great distances.

But these already known facts are not the most important; we mention them only to recall them and put the whole into our own work. The fossils enclosed in this mass, and those contained in the marl that overlies it, offer

24. [The authors describe the three "masses" in the standard geognostic order, from bottom to top, whereas the quarrymen's numbering follows the order in which they were encountered in quarrying, from top to bottom: hence the somewhat confusing description.]

25. [Desmarest had earlier made a study of the prismatic jointing of volcanic basalts; his paper on the gypsum ("Couches de la colline de Montmartre," 1804) had shown how its prismatic jointing could be explained similarly in terms of shrinkage.]

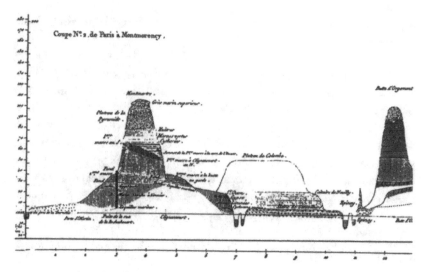

FIGURE 18   A section through Montmartre, part of one of the measured sections that illustrated the full version of the report by Cuvier and Brongniart (1811). The dramatic vertical exaggeration allowed much detail to be depicted in these horizontal beds. The lower part of the hill shows the Gypseous Formation with freshwater shells (e.g. "Lymnées"); above are the marine shells ("Cytherées") that marked the resumption of marine conditions (see fig. 19); and the hill is capped by the Marine Sandstone. The low ground near the Seine (cut three times as it meanders) is occupied by the superficial Detrital Silt, interpreted as dating from *after* whatever event scoured out the valley and left the hill isolated. The vertical scale is in meters above sea level, the baseline is the level of the Seine in Paris, and the horizontal scale is in kilometers north-northwest of the center of Paris.

observations of quite another interest. It is in this first mass that the skeletons of unknown birds and quadrupeds are found daily, which one of us[26] has described in detail in other articles. To the north of Paris, they are in the gypseous mass itself, they have preserved some solidity there, and they are enveloped only by a very thin bed of calcareous marl; but in the quarries to the south, they are often in the marl that separates the gypseous beds, and are then very friable. We shall say no more about the way they are situated within the mass, their state of preservation, their species, etc.; these points have been sufficiently developed in the memoirs we have just mentioned. Some turtle bones and fish skeletons have also been found in this mass.

26. Mr. Cuvier, *Annals of the Museum of Natural History,* vol. __. [The reference was left blank in both published texts; Cuvier had in fact been publishing articles on the fossils from the gypsum since volume 3 (1804).]

What is more remarkable, however, and far more important for the consequences that result from it, is that freshwater shells are found in it, although very rarely. Anyway, [even] a single one is sufficient to demonstrate the truth of the opinion of Lamanon and some other naturalists, who think that the gypsums of Montmartre and other hills in the Paris Basin crystallized in freshwater lakes. We shall report in a moment some new facts that confirm this.

Lastly, this upper mass is essentially characterized by the presence of mammal skeletons. These fossil bones serve to identify it when it is isolated, for we have never been able to find them, or to confirm that they have been found, in the lower masses.

Above the gypsum are placed thick beds of marl, sometimes calcareous, sometimes argillaceous. In the lower strata and in a white, friable, calcareous marl, the trunks of palm trees petrified in silica have been found on several occasions. They are lying flat and are of considerable volume. In the same set of beds—but only at Romainville—shells have been found of the genera *Lymnaea* and *Planorbis*, which appear to differ in no way from the species that live in our marls. One of us has already reported this interesting fact to the class.[27] It proves that these marls are of freshwater formation, like the gypsums they overlie.

Above these white marls, very numerous and often massive beds of argillaceous or calcareous marls are seen again. No fossils have yet been found in them.

One then finds a little bed, six decimeters thick, of a yellowish foliated marl. In its lower part it encloses some nodules of earthy strontium sulfate; and a little higher, a thin bed of small elongated tellinas lying packed against each other [fig. 19].[28] This stratum, which seems to be of very little importance, is [in fact] remarkable, first by its great extent: we have observed it over an area of more than ten leagues [30 miles] in length and more than four in breadth, always in the same place and with the same thickness. It is so thin that one has to know exactly where to look in order to find it. Second, because it serves as the limit of the freshwater formation, and indicates the sudden start of a new marine formation.[29] In fact all the

27. [Brongniart's paper "Terrains formés sous l'eau douce" was eventually published in 1810.]

28. [Shells of the bivalve genus *Tellina* had already been listed among many other *marine* mollusks in the Coarse Limestone; their importance was that they were taken to mark the resumption of marine conditions, after the freshwater conditions of the rest of the gypsum formation (see below). By the time the full report was published, the authors had assigned the shells to another marine genus, *Cytherea*, and named them "cytherée bombée."]

29. [The crucial word "sudden" *(subit)* is not in the original manuscript, and must have been inserted at a late stage, while the article was in press. As usual, no special evidence is cited to support the suddenness of the change.]

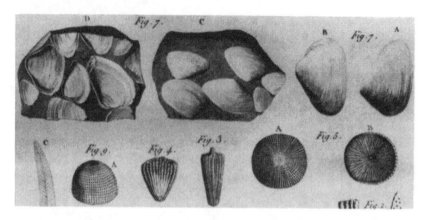

FIGURE 19 Significant fossils from the formations of the Paris region, illustrating the full report by Cuvier and Brongniart (1811). In the upper row are the small shells found immediately above the gypsum beds, which the authors took as evidence for a sudden reversion to marine conditions. In the lower row are small corals from the Coarse Limestone, which—since living corals are intolerant of even slightly brackish water—confirmed the authors' inference that that formation had been deposited during an earlier marine phase.

shells that one comes across above this stratum of tellinas are, like them, marine.

First, and immediately following, a prominent and constant bed of greenish argillaceous marl is found; it can be recognized from afar by its thickness, its color, and its continuity. It acts as a guide to locate the tellinas, since they are found below it. Moreover it contains no fossils, but only argillo-calcareous geodes and nodules of strontium sulfate. It is used for making coarse earthenware.

The four or five beds of marl that follow the green marls are thin and likewise do not appear to contain any fossils; but these strata are immediately covered by a bed of yellow argillaceous marl that is molded by the debris of marine shellfish, the species of which belong to the genera Cerithium, Trochus, Mactra, Venus, Cardium, etc. Fragments of the palate of a ray that must be related to the eagle ray are also found in it.

The marl beds that follow that one show almost all the marine fossil shells, but only the bivalves; and the last beds, which are immediately underneath the argillaceous sand, enclose two distinct beds of oysters. The first and lower is composed of very thick, large oysters: some are more than a decimeter long. Then comes a bed of whitish marl without shells, and then a second, very massive bed of oysters, subdivided into several beds. These oysters are brown, much smaller, and much thinner than the preceding ones. These last beds of oysters are very constant, and we have perhaps not found them missing [even] twice in the many hills of gypsum we have examined. The gypseous formation is often finished off with a more or less thick mass of argillaceous sand that encloses no shells.

Such are the beds that generally make up the gypseous formation. We were tempted to divide it in two, and to separate the description of the marine marls at the top from those of the gypsum and of the freshwater marls at the bottom. But the beds are so similar to each other, and accompany each other so constantly, that we thought we should be content to point out this division without making it in reality.

It remains to say a few words on the principal differences among the hills that belong to this formation. The gypseous hills form a kind of long, broad belt oriented from southeast to northwest, with a breadth of about six leagues. It appears that in this zone only the central hills show the three masses of gypsum distinctly. Those at the edges, such as the plaster quarries of Clamart, Bagneux, Antony, Mont-Valérien, Grisy, etc., and those at the extremities, such as the quarries at Chelles and Triel, have only one mass. This mass seems to us to be equivalent to the one the quarrymen call the first, that is to say, the uppermost; for the fossil mammals that characterize that mass are found there, whereas in its marls one does not come across the numerous coarse crystals of lenticular gypsum that are seen in the marls of the second and third masses.

Sometimes the upper marls are almost completely missing; sometimes it is the gypsum itself that is totally missing or reduced to a thin bed. In the first case the formation is represented by the green marls accompanied by strontianite. The gypseous formations in the park at Versailles, near Saint-Cyr, and those at Viroflay are in the first category; those of Meudon and Ville-d'Avray are in the second.

We should recall here what one of us has said elsewhere,[30] namely that the gypseous terrain of Paris cannot be correlated exactly with any of the formations described by Mr. Werner or his disciples. We then inferred the reasons for this, which it is pointless to repeat.

5. SAND AND MARINE SANDSTONE FORMATION [GRÈS MARIN]

This terrain is not extensive, and appears to follow the formation of the gypsum marls. We would even have united it with them, had it accompanied them as consistently as they accompany the gypsum, and had it not often been separated by a considerable mass of argillaceous sand devoid of

30. Brongniart, *Elementary treatise on mineralogy* [*Traité élémentaire de minéralogie,* 1807], vol. 1, p. 177 [where he noted that the Paris gypsum lay above the Coarse Limestone; the gypsum formations listed by Werner clearly lay much lower down the geognostic sequence].

any fossils, very different in nature from that which will [now] concern us.

What we have just said shows that this formation generally covers the gypseous formation. It consists of beds of siliceous sand, often very pure and often cemented into sandstone, which encloses very diverse marine shells, all of the same species as those at Grignon.[31] We have recognized the same oysters, the same calyptraeas, the same tellinas, the same cerithia. Sometimes these shells still exist in a calcareous state, sometimes only the imprints or external molds remain.

These sandstones and marine sands are found at the summit of Montmartre, at Romainville, at Saint-Prix, near Montmorency, at Longjumeau, etc. At the latter, fossil barnacles may be noted.

Observing these sandstones filled with the same shells as at Grignon, one cannot avoid reflecting on the singular circumstances that must have obtained at the formation of the beds we have just examined. In recalling the beds since the chalk, one pictures [se représente] first a sea that deposits on its floor an immense mass of chalk and of mollusks of distinctive species.[32] This precipitation of chalk and of the accompanying shells suddenly ceases. Beds of a quite different nature succeed it, and only clay and sand are deposited, without any organisms. Another sea returns:[33] this one sustains a prodigious quantity of shelled mollusks, wholly different from those of the chalk. Massive beds are formed on its floor, consisting in large part of the shelly coverings of these mollusks; but little by little this production of shells diminishes, and ceases completely. Then the ground is covered with freshwater; and alternating beds of gypsum and marl are formed, which envelop the debris of the animals that these lakes sustained, and the bones of those that lived on its shores. The sea returns a third time, and produces some species of bivalve shells and turbos; but soon this sea gave birth to nothing but oysters. Finally the products of the second sea (below) reappeared, and the same shells are found at the summit of Montmartre as those found at Grignon and in the depths of the quarries at Gentilly and Meudon.

31. [Lamarck's best locality for the fossil shells of the Coarse Limestone lower down; hence, an indication of a return to the *same* conditions as before the freshwater episode of the Gypseous Formation (see below). This shows that Cuvier and Brongniart were using fossils primarily to infer paleoecological conditions, rather than to characterize specific formations as Smith was doing.]

32. [The striking use of the present tense, combined with the verb "se représenter," accentuates the status of this passage as a verbal *reconstruction* of a dynamic sequence of past environments, analogous to Cuvier's reconstructions of his fossils as living animals.]

33. [As noted earlier, the authors assume that the Plastic Clay represented a *freshwater* phase, although they had no positive fossil evidence for this. Their assumption made it analogous to the Gypseous Formation, likewise deposited during a freshwater phase between two marine periods.]

6. SILICEOUS LIMESTONE FORMATION [CALCAIRE SILICIEUX]

The formation we are going to speak about has a geological situation parallel, as it were, to that of the marine limestone.[34] It is situated neither below it, nor above it, but beside it; and it seems to maintain that position over the immense extent of terrain that it covers to the east and southeast of Paris. This terrain is placed immediately above the plastic clays. It is formed of distinct layers of limestone, sometimes soft and white, sometimes gray and compact, and of very fine grain, penetrated with silica that has infiltrated itself in all directions and to every point. Since it is often cavernous, this silica, in infiltrating its cavities, has coated the inner surfaces with knobbly stalactites of various colors, or with quartz crystals that are very squat, almost without prisms, but clear and limpid. This character is very noticeable at Champigny. This compact limestone, impregnated thus with silica, yields on burning a lime of very high quality.

But the distinctive character of this unique formation—this formation that no one had noticed before us, although it covers a considerable extent of terrain—is that it contains no fossils, either marine or fluviatile; at least, we have not been able to discover any in the large number of places where we have examined it with the most scrupulous attention.

It is in this terrain that the stones known under the name of "millstones" [meulières] are found. The origin, formation, and situation of these stones were obscure to the majority of mineralogists, but they seem to be the siliceous frame of the siliceous limestone. Deprived of its calcareous part by some unknown cause, the silica would have remained, and in effect leaves porous but hard masses; their cavities still contain the argillaceous marl, and they show no trace of stratification. We have made genuine artificial "millstones" by throwing the siliceous limestone into nitric acid. In the second part [of this paper] we shall describe the various areas that are formed from this limestone. We will end this general description of it by saying that it is often exposed at the surface of the ground; but also often covered by argillaceous marls, by sandstone without shells, and finally by freshwater terrain.[35] Such is the structure of the ground in the Forest of Fontainebleau.

34. [That is, the Coarse Limestone (sec. 3 above). The authors here depart from a simple enumeration of the formations in their geognostic (stratigraphical) order, and revert to a lower or earlier point in the sequence; they also abandon any simple sequence, since they claim that this formation is—as later geologists would put it—the *lateral equivalent* of the Coarse Limestone (see fig. 16).]

35. [Of these three formations, which effectively determine the upper limit of the Siliceous Limestone within the sequence, the last two are those described immediately below. The authors may have regarded the "argillaceous marls" as the equivalent of the Gypseous Formation; if so, the Siliceous Limestone would be clearly located as the lateral equivalent of the Coarse Limestone.]

7. SANDSTONE WITHOUT SHELLS FORMATION [GRÈS SANS COQUILLE]

The sandstone without shells, wherever it is found, is always the last formation or the last but one. It constantly overlies the others, and is never covered by any but the formation of the freshwater terrain. Its beds are often
very thick, and intermingled with beds of sand of the same nature. The
sand that underlies the upper beds has sometimes been swept away by water; the beds are then broken up and have rolled down the flanks of the
hills that they form. Such are the sandstones in the Forest of Fontainebleau, those at Palaiseau, etc.

Not only do this sandstone and sand contain no fossils at all, but they are
often very pure and provide sands valued in the [practical] arts, which can
be collected at Étampes and Fontainebleau, in the Aumont butte, etc.

However, they are sometimes either modified by a blending with clay, or
colored by iron oxides, or impregnated with carbonate of lime that penetrated them by infiltration when they were covered by the calcareous freshwater terrain; such is again the case with the sandstones in several parts of
the Forest of Fontainebleau.

8. FRESHWATER FORMATION [TERRAIN D'EAU DOUCE]

This formation consistently covers all the others. In some respects, in structure and other external characters, the rock that has resulted resembles the
siliceous limestone; that is to say, it is sometimes compact, sometimes soft
and white, but almost always penetrated by a siliceous infiltration. The silica itself, sometimes opaque and yellowish, sometimes brown and translu-
cent like *pyromaque* silica, sometimes replaces the limestone completely;
finally this formation, like the sixth [i.e. the siliceous limestone], yields
millstones that owe their origin to the same cause.

What characterizes this formation uniquely, then, is on the one hand
the presence of shells that are clearly from freshwater, and wholly similar to those we find in our marshes. These shells are three species of Lymnaea and some Planorbis. Also found in this formation are some little
round fluted bodies, which Mr. Lamarck named *Gyroconites*. Their living
analogue is no longer known, but their position shows us that the organism
of which they were a part lived in freshwater.[36]

The second character of this formation is the facility with which the
limestone that composes it mixes with water, however hard it appears to be

36. [Brongniart gave a full description of them in his paper "Terrains formés sous l'eau douce"
(1810); soon afterward, Desmarest's son Anselme-Gaétan (1784–1838) identified them as the calcified
fruits of the freshwater plant *Chara* ("Sur la Gyroconite," 1811), thus confirming the interpretation
of the formation as freshwater in origin.]

at the time it is extracted from the quarry: hence the considerable use that is made of it as fertilizing marl at Trappe, near Versailles, in the plain of Gonesse, and throughout the Beauce.

We include in this formation, though with a little uncertainty, the hilltop sands that enclose wood and parts of plants changed to silica. We have been led to make this connection, by observing the silicified wood and plants that are found toward the summit of the hills at Longjumeau. The same sand that contains these plants also contains flints filled with large Lymnaea and with Planorbis.

The freshwater formation, although always superficial, is found in all kinds of situations, toward the summits of hills and on broad plateaus, however, rather than on the floors of valleys. Where it exists in the latter, it has been covered by the material that constitutes the ninth and last formation. Moreover, it is extremely common everywhere in the environs of Paris, and probably at much greater distances than those to which we have been. In view of that, it seems surprising to us that so few naturalists have paid attention to it: we only know of Mr. Coupé who has mentioned it.

The presence of this formation implies that the freshwater that then existed had properties that we no longer find in those we know today. The waters of our marshes, lagoons, and lakes deposit only friable silt. In none of them have we observed the properties that the freshwater of the former world possessed, of forming thick deposits of hard yellowish limestone, white marls, and often very homogeneous flint, enveloping all the remains of the organisms that lived in these waters, and even giving them the siliceous and calcareous nature of their matrix.

9. DETRITAL SILT FORMATION [LIMON D'ATTERRISSEMENT]

Not knowing how to designate this formation, we have given it the name of silt, which indicates a mixture of materials deposited by freshwater. In effect, the detrital silt is composed of sands of all colors, of marls and clays, and even of a mixture of those three materials impregnated with carbon, which gives it a brown or even black appearance. It contains smooth pebbles; but what characterize it more particularly are the remains of the large organisms that one observes in it. It is in this formation that large tree trunks and the bones of elephants, oxen, antelopes, and other large mammals are found. Also belonging to this formation are the deposits of smooth pebbles on valley floors, and probably also those on some plateaus, such as the Bois de Boulogne, the plain between Nanterre and Châtou, certain parts of the Forest of Saint-Germain, etc.

The detrital silt is found not only on the floors of now existing valleys;

it [also] covered valleys or hollows that have since filled up. This arrangement can be seen in the deep trench made near Séran to carry the Ourcq Canal. This trench has revealed a section through an ancient cavity filled with the materials that compose the detrital silt; and in this kind of marsh bed the bones of elephants and large tree trunks have been found.

It is to the existence of these remains of organisms, not yet completely decomposed, that one must attribute the dangerous and even pestilential emanations that are released from these soils, when they are disturbed for the first time after the long succession of centuries that has elapsed since their deposition. For this formation, which appears so modern, is like all those we have just been examining. Although very modern in comparison to the others, it is still anterior to historical times; one could say that the silt of the ancient world *[ancien monde]* in no way resembles that of the present world *[monde actuel]*, since the wood and animals found there are entirely different, not only from the animals of the countries in which they are found deposited, but even from all those that are hitherto known [anywhere].

[CONCLUSION]

We have just given an outline *[tableau]* of the nature, structure, and particularities that characterize the different systems of beds that compose the formations of the Paris region. We have determined the order of their superposition and consequently the order of their deposition. We have still to describe the districts that they form or cover; this will be the object of a second paper.[37]

Translated from Cuvier and Brongniart, "Géographie minéralogique des environs de Paris" (1808), with additions from the manuscript of the original text, MS 631, Bibliothèque Centrale, Muséum National d'Histoire Naturelle, Paris.

37. [This, the concluding paragraph of the manuscript, was omitted from the paper as printed in 1808; the promised further paper, with a mass of local details, was published as a long "second chapter" in the full version of 1811, and introduced with a passage adapted from this paragraph.]

13

FOSSIL DEER AND CATTLE

———————

Meanwhile Cuvier's papers on specific fossils continued to appear in every volume of the *Annales du Muséum*. The series on the palaeotherium and anoplotherium tailed off without the conclusions and reconstructions that he had drafted (text 7); but he added to the Parisian fauna with descriptions of the bones of birds, turtles, and a fox-sized carnivore.

Cuvier also began to publish studies of fossils that now—thanks to his recent survey of the geological literature (text 14)—he knew were distinctly older even than the Parisian ones. For example, studies of the osteology of living crocodiles gave him an authoritative basis on which to claim that certain fossils from Normandy and Thuringia were crocodiles and monitors, though of course distinct from the modern forms. These reptiles, he emphasized, came from "very ancient Secondary beds," far older than the Parisian formations.[1] Another paper dealt with a unique specimen that Scheuchzer, almost a century earlier, had famously claimed to be the skeleton of "a man who was a witness of the Deluge" (*Homo diluvii testis*, 1726). Cuvier, in a debunking gesture worthy of his Enlightenment forebears, demonstrated that it had been a giant amphibian

1. In modern terms the fossils from Normandy were Jurassic in age; those from Thuringia, Permian. Historically, the important point is that neither Cuvier nor any of his contemporaries were yet certain about their relative ages, and the standard terms for the major periods did not come into general use until after Cuvier's death.

of the salamander group! He also described fully, though only from a published illustration, a unique specimen of what he termed a "pterodactyle," from the lithographic stone of Solnhofen in Bavaria. Following his preliminary assessment of it (text 6), he interpreted it as a *flying* reptile: that this was a striking conclusion should need no emphasis.

On Werner's authority, Cuvier believed that the rocks from which some of these fossils came were among the oldest of the Secondary formations. That the fossils were all reptilian, not mammalian, strengthened Cuvier's growing suspicion that an age of reptiles had preceded that of mammals in the history of life. These studies of reptilian fossils also provided the occasion for Cuvier to launch a devastating attack on Faujas: not this time on his colleague's geology, but on what Faujas had had the temerity to claim in Cuvier's own field of comparative anatomy.

The most celebrated of all the vertebrate fossils from the older formations was the so-called Maastricht animal. Its large bones had been found from time to time since the mid–eighteenth century, in the underground Chalk quarries at Maastricht in the Netherlands. The most spectacular specimen, like the elephant skulls that had first given Cuvier the key to the mammoth problem (text 3), had been one of the trophies of the revolutionary wars, and was now in the Muséum's collections. Indeed it was Faujas himself who had brought it to Paris; and he had later made the huge jaws the centerpiece of his massive monograph on the fossils from Maastricht (*Montagne de Saint-Pierre de Maestricht*, 1799).

Like the equally celebrated "Ohio animal," the affinities of the "Maastricht animal" were controversial. The Dutch anatomist Petrus Camper had thought it a toothed whale, but Faujas had claimed it was a crocodile. Camper's son Adriaan, who had inherited his father's great anatomical collection, suggested to Cuvier that it was a giant lizard. Cuvier at first rejected that startling idea, but later he adopted it: a rare occasion on which he deferred to another naturalist. In his published paper he scornfully rejected Faujas's attempt at comparative anatomy, and concluded that the animal had been a marine lizard of the monitor group: that interpretation was built into the name he gave the animal, *Mosasaurus*, or "lizard from the river Meuse."[2] Its monstrous size was no more strange, he pointed out, than that of the elephant-sized megatherium and the rhinoceros-sized megalonyx he had already described. Not only did Cuvier's conclusion crush Faujas's pretensions as a fossil anatomist; by also rejecting the elder Camper's suggestion that the "Maastricht animal" was

2. Cuvier, "Grande animale fossile de Maestricht" (1808). Maastricht is on the river Maas or Meuse.

a whale, and agreeing with the younger that it was a reptile, Cuvier reinforced his view that the older Secondary formations represented an age without mammals.

With these fossil reptiles, Cuvier pressed the history of the vertebrates as far back as he had empirical material to go: to have gone any further would have been to indulge in the kind of conjecture he deplored. At the other end of the history of life, however, he remained acutely aware of the need to define more precisely what had happened at the last and most recent "revolution."

One category of material relevant to that problem, on which Cuvier published several papers around this time, was that of the fossil bones that had long been collected from many caves, particularly in Bavaria and other parts of Germany, and in some cases in extraordinary abundance. He concluded that most of the bones had belonged to a species of bear, but to one distinct from the polar bear, the European bear, or any other living species. Other bones included those of a hyena, a lion or tiger, a wolf, and a fox, all likewise distinct from living species. This study put the cave bones unproblematically into the same category as those of mammoths, rhinoceros, and other large mammals from the superficial deposits: all were relics of a fauna that had become wholly extinct—"destroyed" was as usual Cuvier's preferred verb—at the last revolution.

More problematic were the "bone breccias" (*brèches osseuses*) that had likewise long been known from Gibraltar and certain parts of Dalmatia. These were rocks of "stalactitic" material, filling fissures or small caves in much more ancient limestone. The material was usually so hard that the embedded fragments of bone were difficult to extract. But Cuvier managed to identify enough to conclude that most if not all the species were extant, and that many still lived in the same regions. He therefore inferred that, although the breccias were very ancient in terms of human history—he thought none were forming any longer—they were much more recent than the superficial deposits containing the bones of mammoths. In other words, they could be dated *later* than the last revolution, as clearly as the mammoth fauna dated from before that event.

One highly problematic category remained. These were the bones of ruminants such as cattle and deer. On the one hand, some such bones were found in the superficial deposits, along with those of mammoths and other extinct species, yet they seemed indistinguishable from those of living ruminants. On the other hand, at least one species, the celebrated giant deer or "elk," was unquestionably extinct, yet seemed to come from truly recent deposits, namely from the peat bogs of Ireland. Altogether, the fossil ruminants represented a worrying anomaly for Cuvier:

ever since the start of his research, he had claimed there was a sharp distinction between the extinct fauna of the "ancient world" before the most recent revolution, and the extant species of the "present world." He therefore devoted an important paper to reviewing this crucial problem (text 16).

Like any modern scientist in a similar situation, Cuvier successfully made the anomaly go away, at least to his own satisfaction. To be fair, he had good reason to claim that the problem with the ruminants was that distinct species often had no discernible difference in the skeletal parts that could be preserved as fossils. This was no special pleading, for it could easily be demonstrated among living species. It followed that the bones of ruminants found with the extinct mammoth fauna could well have belonged to species that were equally extinct, yet it might be impossible to demonstrate that fact conclusively from their bones. Much less convincingly, Cuvier explained away the Irish "elk," by claiming that its bones came not from the peat itself but from underlying deposits that might date from the time of the last revolution. He therefore argued that the fossil ruminants were no good reason for abandoning the sharp distinction he had inferred from all the less ambiguous evidence. He concluded that the bones of ruminants came from two distinct kinds of deposit and two separate "epochs": either they belonged to animals of the present world—and were therefore not truly fossils at all—or they represented animals that had been as much the victims of the last revolution as the mammoths and other clearly extinct species. However, Cuvier emphasized that his conclusions on this issue were only provisional; his tentative tone was in striking contrast to his posthumous reputation for scientific dogmatism.

TEXT 16

On the Fossil Bones of Ruminants Found in
the Superficial Deposits

By G. Cuvier

THE SPECIES OF RUMINANTS are the most difficult to distinguish from one another. Although they are sharply distinct from other quadrupeds, they resemble each other so much that to characterize genera one has to use parts such as the horns; and these are not only altogether exterior, and consequently of little importance,[3] but also vary—within the same species—in form and size, even to the point of being missing altogether in some circumstances, according to sex, age, and climate.

The difficulties that ruminants pose for geology, however, are even greater—if that is possible—than those involved in distinguishing their bones. In the superficial deposits [terrains meubles], we have found up to now only pachyderms that differ in species from those of today. The carnivores that accompany them are at least very foreign to our climate. The caves themselves offer us hardly anything but unknown or exotic carnivores. But among the ruminants, almost all the species we find as fossils, whether in the superficial deposits or in rock fissures filled with stalactite, seem to differ in no essentials from those of our own country and our own time.

The fossil elk [élan] of Ireland, which seems to be truly lost [perdu], is indeed an exception to the rule, and fits into what we have observed relative to the pachyderms. Some species of deer may yet belong with it, but I must admit it has been impossible for me not to recognize the skulls of aurochs, cattle, and certain bison [buffles] for what they truly are [fig. 20]. The horse genus shares with the ruminants this resemblance between fossil bones and those of living species.

In truth, most of the bones of horse, cattle, and aurochs that I have observed were taken from the most recent alluvial deposits, or even from peat bogs; some also came from sands that could have caved in on them. But there are some that are not in such situations, and the bones of elephants and rhinoceros are scarcely ever found unaccompanied by those of cattle, bison, and horses. There were thousands of them in the famous deposit at Canstatt. I myself have seen them retrieved by the hundreds from the Ourcq Canal, without being able to see any difference between their position [gisement] and that of the elephant bones taken from the same canal.[4]

Did these bones belong to races of which a few individuals, by retreating onto mountains, escaped from the catastrophe that buried the elephants and rhinoceros on our plains? Or have the deposits, in which they are found mixed pell-mell with extinct races, been displaced [remués] after the destruction of the latter? Or indeed were these species of ruminants distinguished from those of today by external characters that can no longer be

---

3. [As usual, and in accordance with his biological principles, Cuvier assumed that external characteristics are always less revealing than internal ones, for determining the true position of vertebrate animals in a natural classification.]

4. [The Ourcq Canal was dug at this time a few miles outside Paris, in the Forest of Bondy; it was the most prolific source of these fossils that was available to Cuvier within easy reach of the Muséum.]

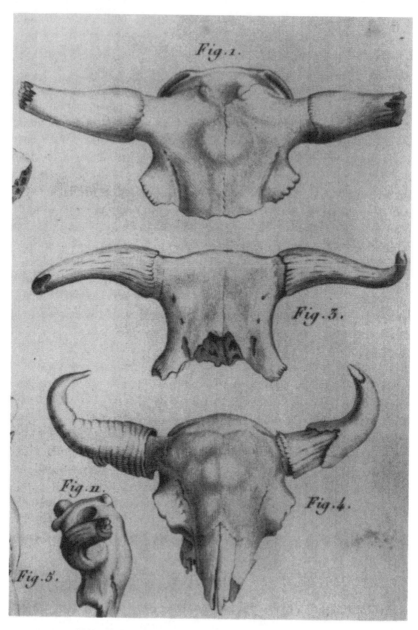

FIGURE 20 Three skulls illustrating Cuvier's paper on fossil ruminants (1809). He noted that all were indistinguishable from living species, except for their larger size; the top figure, for example, was a skull like that of the ancient wild aurochs; the bottom figure, a Siberian skull like that of the European bison.

recovered from their skeletons, as the zebra differs from the ass, for example, or the quagga from the horse? Or else, finally, is it possible that what has been collected along with the bones of elephants and suchlike are only the noncharacteristic parts that were identical in the extinct and the living species, while the skulls and other distinctive parts that are similar to those of living species have been recovered only from modern deposits?

These four cases are all possible. Which actually took place? I do not yet dare decide. Perhaps the continuation of our researches will give us some incentives to be bolder. In the meantime, let us follow the path, and seek to complete the essential objective, which is the identification of the bones. To do so, let us start by setting out in a few words the main osteological characters common to all the ruminants, indicating some of those that can best serve to distinguish the genera.

[Here follow detailed descriptions and analyses of the fossil bones of deer and cattle, based on specimens in the Muséum and on published and unpublished illustrations.]

GENERAL SUMMARY

After this examination, it can be seen that the bones of ruminants in the superficial deposits, as far as it is possible to distinguish them, relate to two classes, as many in the deer genus as in that of cattle: namely, that of the bones of unknown ruminants, in which we place the Irish elk, the small slender-horned deer from Étampes, the deer from Scania, and the large buffalo from Siberia; and that of known ruminants, which are the ordinary [red] deer, the ordinary roe deer, the aurochs (which appears to be the original stock of our domestic cow), and the buffalo with converging horns (which seems the same *[analogue]* as the musk ox of Canada). After that we are left with one doubtful species, namely the large fallow deer of the Somme, which is very similar to the common fallow deer.

The positions *[gisemens]* of all these bones are far from being known exactly; but if those that are known are compared, it is found that the known species are always in deposits *[terrains]* that appear to be more recent than the others. That much is certain, at least for the [red] deer, roe deer, and cattle of the Somme valley, which are in loose and superficial sands or in peats. The aurochs seem likewise to be found always in alluvial deposits *[alluvions]* or recent detritus *[atterrissemens]* that are still able to be accumulated or eroded *[diminués]*; and the antlers of English [red] deer have often been recovered from the beds of rivers themselves.

As for the unknown species, it was noticeable that the Irish elk, although one must go through beds of peat to find it, is not in the peat itself, but rather in the beds of clay or chalk underlying it. The deer of Étampes, found in the sands of the Beauce, was below the freshwater formation that covers the sands. Finally, the Siberian bison [buffle], accompanying fossil elephants and rhinoceros, must be of the same age and in the same beds. Among the unknown species, only the deer from Scania is reported to have been found in a peat bog, but perhaps this claim would be worth verifying.

In view of the little attention that has hitherto been given to the positions of fossil bones, the result I offer is doubtless still very shaky; so I do not claim to assign it value except as an indication that deserves to be examined by naturalists who have the opportunity to do so.

One observation of another kind already has more certainty. The known fossil ruminants are also animals belonging to the climate in which they are found. Thus the [red] deer, the cow, the aurochs, the roe deer, and the musk ox of Canada live and have always lived in cold and temperate countries; whereas the species that we regard as unknown — if one wants, in spite of all opposition, to relate them to their existing analogues — would find those analogues only in warm countries. Our unknown fossil ruminants follow this analogy in part. The large Siberian bison can only be compared with the Indian buffalo or to the water buffalo [arni]; just as it is only in the Indian elephant and the African rhinoceros that it has been alleged that the original mammoth and fossil rhinoceros are seen, with which are found the bones of that [fossil] buffalo.

In truth, the Irish elk and the deer of Étampes and Scania could be compared to animals from cold countries, but they are not close enough to weaken our reasoning. The facts collected up to the present therefore seem to show — at least as far as such incomplete documents can do so — that the two sorts of fossil ruminants belong to two kinds of formation [terrain], and consequently to two different geological epochs: that the first kind were buried and are still buried daily in the period in which we are living; while the other kind were victims of the same revolution that has destroyed the other fossils of the superficial deposits, such as the mammoths, mastodons, and all the pachyderms of which the genera live today only in the tropical zone.

Translated from Cuvier, "Os fossiles de ruminans" (Fossil bones of ruminants, 1809), omitting some passages of osteological description.

14

COLLECTED RESEARCHES ON

FOSSIL BONES

B y 1810 the spate of Cuvier's papers on fossil vertebrates in the *Annales du Muséum* had virtually dried up. In that and the following year, he contributed only a single minor paper (on reptile and fish bones from the Paris rocks). However, Cuvier did not fade from the scientific scene or from the public eye. His and Delambre's report on the progress of the sciences (text 14) was finally published in 1810, after many delays; and the revised and greatly expanded version of his and Brongniart's study of the "mineral geography" of the Paris region (text 15) was likewise finally published by the Institut in 1811.

Among other reasons for the decline in Cuvier's output of original work, his new appointment to a major administrative position in higher education was certainly important. It was his responsibility to supervise the incorporation of the universities of the newer territories of Napoleon's empire into the reorganized French system. This work took him to Italy, and then to the Netherlands and Germany; these were his first travels outside France since the start of his career. He made good use of his spare time to inspect fossil collections at first hand, to build up his contacts with foreign informants, and—not least—to convince them that their collaboration would not lead to the enforced removal of their collections to Paris.

In effect, however, Cuvier's fossil work was now virtually complete. Although new collections and new finds would amplify the material at his disposal, he had published papers on all the main categories of bones that were known, from all parts of the world, within the limits he had set himself. One of his last papers on fossils to be published in the *Annales,* on seals and sea cows, made those limits clear almost in passing. He explained that such *marine* mammals were really outside the scope of his research, because unlike terrestrial or even freshwater quadrupeds they would not have been destroyed by any marine inundation. That, by implication, had been the character of the last revolution; and its effects, again by implication, remained the center of his interests.

Cuvier had virtually completed his fossil research; it remained to make it more widely available. The volumes of the *Annales* were subscribed to, or received in exchange, by all major scientific libraries; but right from the start Cuvier had planned to reissue his papers in collected form. As they were published, he arranged for a substantial number of extra copies to be printed off, unchanged apart from being separately paginated. Some of these he used as offprints to distribute at once to deserving informants; but a good stock was held in reserve. In 1812 they were finally bound up and reissued in the volumes of his *Recherches sur les ossemens fossiles* (Researches on fossil bones) (fig. 21).

Cuvier took the opportunity to rearrange the papers in a more logical order than that in which they had appeared in the *Annales.* One volume was devoted to all the pachyderms from the superficial deposits; the next to all the fossils from the Parisian rocks; the last was more miscellaneous, but included collections on the ruminants, the carnivores, and the reptiles. Those three volumes were originally all that Cuvier planned to publish.

When *Ossemens fossiles* appeared, however, it was prefixed with an extra volume. Partly it was to house the reissue of his and Brongniart's monograph on the geology of the Paris region: this was now inflated to almost book length on its own, and was too large to put in the same volume as the papers on the Paris fossils, as he had earlier intended. But partly the extra volume was to accommodate a new and lengthy "Discours préliminaire" (Preliminary discourse, text 19), which aimed to set the detailed papers in a general context of significance; perhaps it was also designed to make the whole set of volumes more attractive to the fossil collectors and other "amateurs" whom Cuvier must have hoped would buy it.

The four-volume work was dedicated to Laplace (text 17). Authors at this period took great care in the wording of their dedications, and above all in the choice of the persons to whom they were addressed. Cuvier's

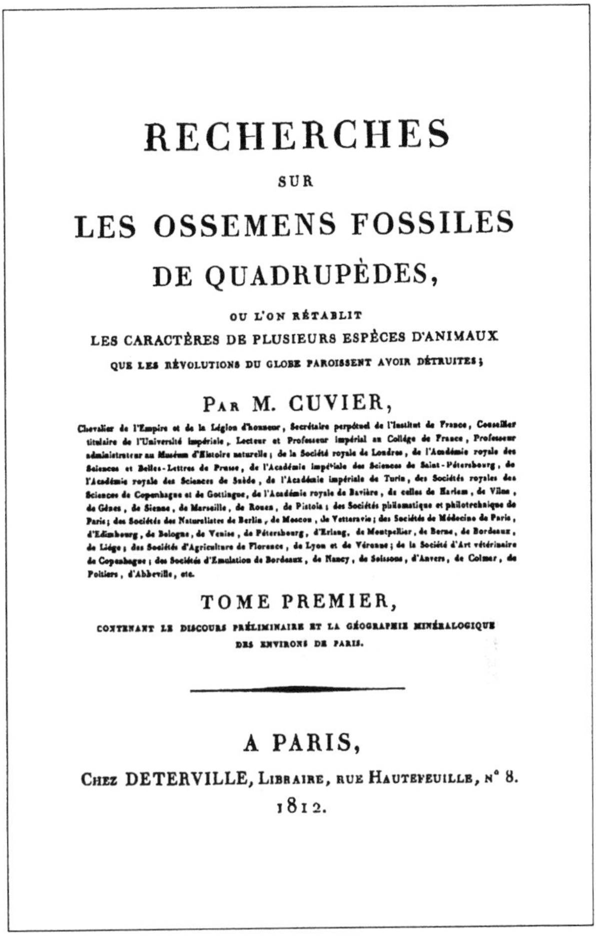

RECHERCHES

SUR

LES OSSEMENS FOSSILES

DE QUADRUPÈDES,

OU L'ON RÉTABLIT

LES CARACTÈRES DE PLUSIEURS ESPÈCES D'ANIMAUX

QUE LES RÉVOLUTIONS DU GLOBE PAROISSENT AVOIR DÉTRUITES;

Par M. CUVIER,

Chevalier de l'Empire et de la Légion d'honneur, Secrétaire perpétuel de l'Institut de France, Conseiller
titulaire de l'Université impériale, Lecteur et Professeur impérial au Collége de France, Professeur
administrateur au Muséum d'Histoire naturelle; de la Société royale de Londres, de l'Académie royale des
Sciences et Belles-Lettres de Prusse, de l'Académie impériale des Sciences de Saint-Pétersbourg, de
l'Académie royale des Sciences de Suède, de l'Académie impériale de Turin, des Sociétés royales des
Sciences de Copenhague et de Gottingue, de l'Académie royale de Bavière, de celles de Harlem, de Vilna,
de Gênes, de Sienne, de Marseille, de Rouen, de Pistoia; des Sociétés philomatique et philotechnique de
Paris; des Sociétés des Naturalistes de Berlin, de Moscou, de Vetteravie; des Sociétés de Médecine de Paris,
d'Edimbourg, de Bologne, de Venise, de Pétersbourg, d'Erlang, de Montpellier, de Berne, de Bordeaux,
de Liége; des Sociétés d'Agriculture de Florence, de Lyon et de Vérone; de la Société d'Art vétérinaire
de Copenhague; des Sociétés d'Emulation de Bordeaux, de Nancy, de Soissons, d'Anvers, de Colmar, de
Poitiers, d'Abbeville, etc.

TOME PREMIER,

CONTENANT LE DISCOURS PRÉLIMINAIRE ET LA GÉOGRAPHIE MINÉRALOGIQUE
DES ENVIRONS DE PARIS.

A PARIS,

CHEZ DETERVILLE, LIBRAIRE, RUE HAUTEFEUILLE, N° 8.

1812.

FIGURE 21   The title page of Cuvier's *Ossemens fossiles* (1812), in which he reissued all the separate papers he had published since 1804. Under his name are listed not only his various positions and honors in Paris, but also the many scientific societies and academies elsewhere to which he had been elected: they reflect his reputation throughout Europe, while serving as a tacit claim to scientific authority.

work was no exception. Pierre Simon, marquis de Laplace (1749–1827), was perhaps the most powerful and prestigious scientific figure in Napoleonic France. His treatise *Méchanique céleste* (Celestial mechanics, 1799–1805) was regarded as having finally perfected the mathematical analysis of planetary motions under universal gravitation, transcending or at least completing the work of the great Isaac Newton more than a century before; his strictly mathematical work, and his rigorous research on physical phenomena such as heat, light, and capillarity were considered equally impressive. By dedicating his own work to Laplace, Cuvier was explicitly presenting it as a project that aspired to similar scientific rigor; he was associating "naturalists" such as himself with the high standards of his "mathematician" colleagues at the Institut. But the flow of prestige was not all one way, as Laplace would have appreciated when he accepted the dedication. If Cuvier's work was acclaimed in its own sphere, Laplace would gain the credit of being associated with research that achieved a rigorous understanding of the natural world in the dimension of geological time, just as his own work had in the dimension of celestial space.

Prefaces too were composed at this period with great care, and here again Cuvier's was no exception (text 18). In it he explained the relation between his separate papers and the collection now being published; and he alluded to two distinct kinds of criticism that his work had already met. The first referred to the continuing debate over the transformist or evolutionary interpretation of species. Cuvier's claims that even the most recent fossils were sharply distinct from any living species, and that no gradual change could have transformed the fossil species into the living, continued to be rejected by some naturalists. In particular, Lamarck had recently made his opposition to Cuvier more explicit than ever before, in his major work *Philosophie zoologique* (Zoological philosophy, 1809), which was devoted to expounding at length his transformist view of life. The other criticism probably referred to Cuvier's claims about the relatively recent date and catastrophic character of the last "revolution"; he stated, however, that he had a much less strong attachment to these claims, and maintained that he would readily abandon them if presented with better explanations.

TEXT 17

⸺

To the Count de Laplace, Grand Officer of the Legion of Honor, Chancellor of the Senate, Member of the [National] Institute and of the Bureau of Longitudes, etc.

MY DEAR AND ILLUSTRIOUS COLLEAGUE,

It is by many rights [titres] that this work is offered to you. When I was still young, and I told you about my first ideas for it, you urged me to follow them. Having since been admitted to sit beside my masters in the Scientific

Class of the Institute, I found advice, encouragement, and help of all kinds. There, above all, I was able to enter into that rigorous spirit *[esprit sévère]* that is the fruit of the felicitous association established in its midst between mathematicians and naturalists. You, Sir, who, after completing the submission of the Heavens to geometry, have applied it with so much success to terrestrial phenomena: you contribute more than anyone to the maintenance of that spirit. It is thus a great privilege for my book, to see your name at its head. For all time it will be inestimable for its author to have received publicly this mark of the esteem and friendship of one of the finest geniuses of his century.

Cuvier
The Botanic Garden, 31 October 1812

Translated from "Épître dédicatoire" (Letter of dedication), in Cuvier, Ossemens fossiles (1812), vol. 1.

TEXT 18

———

Preface

THE AUTHOR DECIDED to publish a large part of his researches on fossil bones as separate papers in the *Annals of the Museum of Natural History*, because in this way he could enable the friends of science to enjoy them as soon as he obtained sufficient information on each kind of bone; and so that the singular results that he thus had to communicate to the public could encourage those who possess such objects, or who were in a position to be able to collect them, to support him in his enterprise.

It was pointed out to him, however, that it could also be useful to make a separate collection of these researches, in which they would be arranged in a methodical order: both for the use of those who do not possess the complete collection of the *Annals*, and also for those who, while possessing this voluminous collection, would be pleased to have all that concerns a kind of fact that is so interesting for the theory of the earth, assembled in a form that is convenient and easy to consult.

Consequently, as these papers were in press, a certain number of copies was printed, which have been bound according to the sequence of the animal families to which they relate.[1] The author has added, in several [new]

---

1. This is why the volumes could not be paginated successively; but as far as possible, tables of contents make up for that fact.

articles and numerous supplementary plates, objects that he has collected since the papers were drafted. At the beginning and the end of each volume he has placed introductions and summaries, in which he presents his principal results under a single viewpoint. At the head of the whole work he has placed a preliminary discourse [text 19], in which he lays out the general principles that have guided his research, the foundations that support them, and the consequences that it seems to him to be possible to deduce, about the physical history of the globe. Finally, one can only have somewhat clear notions about the origin of fossil bones, and the catastrophes that have reduced them to that state, when one knows well the beds [couches] that conceal them, those that cover them, those on which they rest, and above all the other animal and plant remains that can fill these three kinds [ordres] of beds.[2] The author has therefore attached to his preliminary discourse a work that, it seems to him, can serve as an example of the method to be followed in the study of beds. This is the work he has done with Mr. Brongniart, on the area around Paris, which in the variety of its beds and the abundance of its fossils is one of the most remarkable regions in Europe.[3]

In these four volumes, therefore, will be found the author's whole series of applications of comparative anatomy to the history of the globe. These applications have seduced him to the point of making him delay by some years the publication of his major work on the first of those sciences.[4] However, he has no grounds to complain about this, for at the same time they have demanded from him research that has enlightened him on several points in his principal work; and perhaps they will have helped to make more generally felt the utility that this work can have. He is now going to dedicate [consacrer] himself directly to it; and during the remainder of his life, all the moments that his obligations [devoirs] leave at his disposal will be employed in the accomplishment of an enterprise toward which his vows [voeux], so to speak, have been directed since his earliest youth.[5]

A few of the author's assertions about the species to which the bones he

2. [A somewhat convoluted way of expressing the importance of knowing the precise geognostic (stratigraphical) positions of the bones.]

3. [The revised and greatly enlarged version of their joint paper of 1808 (text 15) had been published by the Institut the previous year (1811), and was reissued in Cuvier's *Ossemens fossiles*, where it occupies more than half the first volume.]

4. [Cuvier's four volumes of *Le règne animal* (The animal kingdom) were eventually published in 1817, five years after *Ossemens fossiles*. In the following sentence Cuvier calls the former, not the latter, his "principal" work.]

5. [Cuvier's choice of words is strikingly religious in tone. Although *consacrer*, *devoirs*, and *voeux* all have secular meanings (as "devote," "duties," and "desires"), their combination here seems to express Cuvier's strong sense of personal calling or vocation (in terms of his Lutheran upbringing, *Beruf*) to the scientific life.]

has examined belong have been attacked by esteemed savants. He has therefore been obliged—although with great regret—to respond in detail to the arguments raised against him, because they concern facts from which the theories to be derived ought to flow as from so many foundations, and because it was necessary before all else to put beyond reach [of doubt] these charters and diplomas [chartes et diplômes] of the history of the globe.[6] The author hopes he has acquitted himself of this obligation in a manner that reconciles what the importance of these documents demands with the consideration due to the age and merit of those persons who were compelled—if only briefly—to adopt this polemical role so little in accord with their usual habits.[7]

As for objections of another kind that have been directed against the conclusions that the author draws from these facts, it did not seem necessary to respond to them here, since they are only a matter of simple reasoning, of which everyone is the judge. The author in no way clings to these conclusions; indeed they enter into his work only as digressions appropriate to reduce somewhat its monotony; and if anyone can draw better ones, he will be the first to abandon his own.[8]

Anyway, the goal that the author set himself in publishing his papers separately has already been attained in part. Fossil bones have become an object of attention for savants and commendable amateurs, and they are collected with more care than before. The author recently undertook travels in Italy, Holland, and Germany, where he examined those in several museums [cabinets]. He saw and drew a substantial collection, made in the Arno valley by a society formed with the laudable aim of making known everything that concerns that beautiful region.[9] The Prince-Viceroy of Italy has acquired, and placed in the museum he has just erected in Milan, the astonishing collection that Mr. Cortesi has made in the region of Piacenza.[10] The excavation of the docks at Antwerp, and the work for the numerous canals and fine roads that the Emperor [Napoleon] is having

6. [The conventional metaphor, by which fossils were treated as *documents* recording the history of the earth, is here elaborated in a notable way: Cuvier likens fossils specifically to *legal* documents of the kinds that were used to establish claims to property rights, professional qualifications, etc., based on historical precedents or personal achievements.]

7. [A thinly veiled allusion to Lamarck, and probably also to de Lamétherie, both of whom had criticized Cuvier's opposition to transformism and were now in their late sixties.]

8. [The allusion is probably to Cuvier's "digressions" from geology into the field of human chronology, to find textual evidence to support his natural-scientific claims about the geologically—and even humanly—recent date of the last "catastrophe" (see text 19, sec. 36).]

9. [The Society of Valdarno (or Val d'Arno, the upper valley of the Arno near Florence) had been founded in 1809 specifically as a result of the discovery of deposits rich in fossil bones.]

10. [The prince-viceroy of the Napoleonic Kingdom of Italy (1805–14) was Eugène de Beauharnais (1781–1824), Napoleon's stepson. He had recently bought the fossil collection assembled by

made at so many places in his vast dominions, have led to the discovery of a great number of bones, which have been preserved by enlightened engineers. All these riches could furnish materials for a substantial supplement, the appearance of which will not be delayed, if the public deigns to accord sufficient interest to the present volumes to engage in that sequel.[11]

Translated from the "Avertissement" in Cuvier, Ossemens fossiles *(1812), vol. 1, pp. i–vi.*

Giuseppe Cortesi (1760–1838), a lawyer and naturalist in Piacenza, making it the core of a new natural history museum in Milan, modeled on the one in Paris. The collection consisted mainly of superbly preserved shells, somewhat similar to those from around Paris, from the formations on the flanks of the Apennines, but there were also important vertebrate bones.]

11. [The much enlarged second edition of *Ossemens fossiles* began to appear nine years later, in 1821.]

15

THE REVOLUTIONS OF THE GLOBE

————

By far the most important new feature of the collected *Ossemens fossiles* was its "Preliminary discourse" (text 19), which followed immediately after the preface. The format of a "preliminary discourse," and indeed the phrase itself, was well established: at least since the great *Encyclopédie* of the Enlightenment, such an essay had been almost a standard feature of any work in French with pretensions to monumentality. But Cuvier's essay was untypical in one important respect. Whereas the papers that made up the rest of the work had been addressed originally to Cuvier's colleagues and informants, and were still aimed at them in this new format, the "Discourse" was clearly based on Cuvier's earlier lectures to a general audience (chapter 8). Prefixing a readable discourse to a series of specialized papers thus represented a bid for the attention of the general educated public as well as naturalists and other savants. Cuvier's "Discourse" was in fact found more than merely readable: it was recognized at the time as a masterpiece of scientific prose, and it continued to be reprinted throughout the nineteenth century, long after Cuvier's death and long after the further development of geological research had made it obsolete in strictly scientific terms.

The "Discourse" opens with a bold and vivid claim (text 19, sec. 1): the focus was not on the wonders of nature, let alone—as would have been expected in an earlier age—on the wisdom of its Creator, but on the savant himself. Expanding the metaphor of the naturalist as archeologist, which he had used frequently ever since he first publicly outlined his research project (text 5), Cuvier presented himself as a "new species of antiquarian." Some earlier naturalists had sought to use the reasoning of antiquarians to reconstruct the history of the earth: the metaphor of fossils as the "coins" or "monuments" of nature was a commonplace. What Cuvier claimed as novel to his project was specifically his use of a hitherto neglected *kind* of evidence, that of fossil bones.

The antiquarian metaphor was taken further, however, in a way that reveals Cuvier's attitude toward his scientific material. Just as savants were currently wrestling with the problem of deciphering the ancient hieroglyphic inscriptions that Napoleon's expedition had brought back from Egypt, so Cuvier had to decipher what his fossil bones *meant.* Their significance was not self-evident; it had to be "read," as it were, in the language of comparative anatomy, a language that had to be learned like any other. Cuvier here showed that his conception of "facts" in science was far more subtle than that word in its modern usage might suggest.

Cuvier's claims for his own ability to decipher the language of fossil bones were, as usual, far from modest. But by contrast he presented the "Discourse" merely as a contribution to a small part of "the theory of the earth," or to what he had earlier defined as an explanatory "general geology" (text 14). To express the grandeur of that theme, however, he used some of his best purple prose: it was nothing less than to do for the time dimension of the natural world what Newton—and, by implication, Laplace too—had done for the dimension of space.

In astronomy it was commonplace knowledge that the scale of the universe dwarfed the earth, and even the entire solar system, by its almost inconceivable magnitude. The achievement of the mathematical astronomers lay not so much in discovering that magnitude, but rather in showing that the workings of the universe were knowable and intelligible to earthbound human beings: it was in that sense that they had "burst the limits of space." Likewise, although the general public might still have thought of the history of the earth in terms of the few thousand years of traditional chronology, Cuvier and his colleagues already had a vastly expanded conception of the magnitude of geological time. Cuvier himself had guessed casually that even the rather recent Paris fossils were probably "thousands of centuries" old (text 8), so that his tacit estimate of the whole timescale must have been quite literally unimaginable. What he was referring to here was therefore not so much the magnitude of time, but rather the competence of the human mind to *know* about events that took place before there were any human beings present to witness and record them. It was in that sense that geology could aspire to "burst the limits of time": to burst through the limit set by the oldest human records, and to gain reliable knowledge of unrecorded *prehuman* (or at least, preliterate) history.

After outlining the plan of the "Discourse" (sec. 2), Cuvier introduced the notion of the "revolutions" that the earth has suffered during that long prehuman past (secs. 3–7). His purpose was to impress on general readers, to whom the idea might be novel, the sheer magnitude of the changes that have occurred throughout the history of the earth—even before the origin of life itself—as part of the course of nature. Yet although he argued forcibly that *some* of these revolutions have been sudden, the picture he sketched was not one of incessant turbulence or a rapid succession of "catastrophes." On the contrary, he pointed out that the existence of thick formations of evenly bedded strata implied long periods of tranquillity. "Catastrophes" had been only occasional events, violent changes in physical geography that had punctuated an unimaginably long and mainly tranquil history. Nonetheless, Cuvier's focus on catastrophes was as marked as in any of his earlier writings, and his language accentuated their "terrible" character. As usual, he claimed that their suddenness was proved by the kind of fossil evidence he himself had collected; the evidence of the last catastrophe was the clearest of all, precisely because it was the most recent and its effects were therefore most apparent.

To demonstrate that the most recent revolution had indeed been a "catastrophe," Cuvier first had to argue that the modest processes or "causes" that are now active were insufficient to explain its observable effects (secs. 8–17). Right at the start he claimed that his "catastrophist" interpretation—to use the term that was applied to it much later—was a minority position. Far from being a scientific reactionary, Cuvier presented himself as an innovator in geological theory, whose research had shown the inadequacy of the explanations offered by many of his predecessors and contemporaries. At this point, he argued, the analogy with human history failed: while the past workings of society could indeed be understood by reference to a constant underlying human nature, the present was *not* after all a wholly adequate key to the history of the earth.

Cuvier's claim that "the thread of operations is broken" (sec. 8) has given rise to more argument among historians than perhaps any other phrase in his published work. But in context, and particularly in the light of his earlier writings, its meaning is not obscure. None of the processes "that still operate," Cuvier argued, is adequate to account for the observed effects; therefore those effects must be attributed to causes of another kind. As usual, he declined to suggest what those other causes might have been; but there is certainly no reason to suppose that he had nonnatural causes in mind for them, any more than for the puzzling long-term drop in the level of the world's oceans.

In retrospect, of course, Cuvier can be seen to have made this argument easy for himself by presenting such a limited range of currently active processes.[1] To be fair, however, his location in a museum and his limited experience of geological fieldwork gave him little reason to think his evaluation of current processes was understated. Anyway, his list was quite conventional, and simply borrowed from contemporary works on the subject; it is similar to what he had summarized in his general report on the sciences (text 14).

Having disposed of currently active processes on earth, those in the heavens could be dismissed quite briefly (sec. 18): no slowly acting astronomical change could possibly account for a sudden catastrophe on earth.[2] Here Cuvier got to the heart of his argument with the purveyors of geological "systems": it was, he claimed, the evident inadequacy of present causes—either terrestrial or celestial—that had forced them into a morass of unfounded speculations and thereby brought the whole science into disrepute.

Cuvier therefore set out a review of those "systems," reaching back a century and a half into the past (secs. 19–21). Beginning with Thomas Burnet, in effect the founder of the genre of "theory of the earth," Cuvier swept forward past Buffon to his own contemporaries. Among the latter, Lamarck was prominent, and here Cuvier revealingly set his antipathy to transformist theories in the context of his colleague's broader ambitions for reconstituting all the sciences. But even those theorists who—unlike Lamarck—had kept their speculations within the established principles of physics and chemistry had still produced a fruitless diversity of explanations. The reason, Cuvier suggested, was simple but fundamental: so few of the relevant empirical conditions were firmly established, that the theorizing was virtually unconstrained. The "theory of the earth" was, in modern terms, grossly underdetermined by the available evidence.

So Cuvier laid out an agenda for geological research, which would serve to constrain the conditions for high-level theorizing (secs. 22–25). As in his report on André's book (text 13), he focused on the need to un-

---

1. This was exactly the line of argument pursued, at the very end of Cuvier's life, by Charles Lyell. Lyell's insistence on the total adequacy of "causes now in operation" would have been regarded by Cuvier not as a startling innovation but as a *revival*—and a major strengthening—of an old and familiar argument.

2. A certain circularity of argument will be apparent here, since the sudden and catastrophic nature of the events was just what was in question.

derstand the fossil-bearing Secondary formations that the practitioners of *Geognosie* had tended to neglect. He pointed out that the whole field had hitherto been cultivated by two separate groups of naturalists: the mere "cabinet naturalists" who disdained fieldwork and just constructed high-level theories indoors, and the mere "mineralogists" who in their otherwise admirable fieldwork neglected the crucial evidence of fossils. Clearly his own program of research was to bring the two together. Likewise he claimed that even fossils themselves had been studied only as objects of curiosity, and not in relation to the rock formations in which they were found; above all, they had not been treated as "historical documents."

Cuvier therefore argued next for the supreme importance of fossils for constructing the "theory of the earth" on firmer foundations (secs. 26, 27). Unlike many of the theorists he criticized, Cuvier was ready to concede his own—and their—utter ignorance about major aspects of the earth's history; but that only served to highlight the strategic significance of his own chosen fossils. Unlike the abundant fossils of marine invertebrates studied by other naturalists, such as Lamarck, the much rarer and more problematic fossil remains of terrestrial vertebrates could, paradoxically, provide more decisive evidence. This, Cuvier claimed, was because the species still living, particularly the larger ones, were relatively well known, and so could provide a reliable baseline for comparison with fossil species. By contrast, naturalists could not have the same degree of certainty about the relation between living and fossil species of marine invertebrates, since knowledge of exotic and—especially—of deepwater faunas was still so defective.[3]

To make this argument convincing, Cuvier next had to discuss the likelihood that species of large living vertebrates still remained to be discovered on land (sec. 28). Here he had to contend with the obvious fact that in his day the interiors of the continents (apart from Europe) were far from being well known, so that his critics could argue that herds of mammoths, mastodons, and megatheriums might still be roaming unexplored areas. He minimized this problem by claiming that long human occupation of even the margins of the continents would be likely, in the course of time, to yield knowledge of all the larger animals of the interior.

3. A striking instance of this had been reported by Lamarck in the *Annales du Muséum,* just at the time that Cuvier was publishing his own papers on vertebrate fossils. A French expedition had brought back from Australia what Lamarck identified as the shells of a living species of *Trigonia,* a distinctive mollusk that until then had been known only as a fossil from relatively ancient Secondary formations (in modern terms, Jurassic) (Lamarck, "Nouvelle espèce de Trigonie" [1804]). The report was as surprising at the time as, for example, the discovery of living coelacanth fish has been in twentieth-century biology.

The most effective way to demonstrate the validity of that argument in turn was to show that all the animals reported or described in the course of human history could be identified with species known at the present day. So Cuvier plunged at this point into an extended review of those known to the writers of classical antiquity, including those that appeared to be merely fabulous. He deployed a formidable knowledge of ancient literature—a product of his broad education, but doubtless amplified by his earlier work for the French edition of *Asiatick researches*— to show that all the creatures mentioned could be assigned to one or another of known living species. He therefore concluded that the large fossil mammals described in detail in his work were very unlikely to survive anywhere; their disappearance must be the result of "general causes." It is important to note that this review entailed giving a highly critical evaluation of reports that were likely to be confusingly garbled, but which might yet contain a kernel of scientific truth: in other words, Cuvier was acting here as a rigorous textual critic.

Having established the strategic advantages of his own research material, for putting the "theory of the earth" on firmer foundations, Cuvier next conceded the practical difficulties of dealing with fossil vertebrates (sec. 29). Unlike, for example, the well-preserved fossil mollusk shells that Lamarck was describing, the scattered bones of fossil vertebrates usually had to be pieced together to reconstruct a skeleton. In fact, however, Cuvier turned this problem to his own advantage, because its successful solution depended on the application of the zoological principles that underlay his main research field of comparative anatomy. Those principles of the functional integration of the animal organism were therefore set out here, perhaps more clearly than anywhere else in Cuvier's work, embedded rather surprisingly in a primarily geological argument (sec. 30). But in addition to such theoretical issues, he also emphasized the crucial practical advantage he enjoyed by working at the Muséum: only there could he have immediate access to the vast range of comparative material needed to identify, or establish the affinities of, fragmentary fossil bones.

Cuvier now summarized briefly the results he had obtained, and the fossil species he had been able to identify (sec. 31), and immediately set out their relation to the succession of rock formations in which they were found (sec. 32). He claimed, in effect, that they demonstrated a true *history* of the vertebrates, in which reptiles had preceded mammals, and aquatic mammals had preceded those adapted to life on land. As usual, his terms for the origins of these major groups were carefully descriptive rather than causal: the reptiles, for example, simply "began to exist" at a certain period. In the younger formations, mammals of genera unknown alive, such as those from the Paris strata, had in turn preceded those that differed only at the specific level, namely the fauna of mammoths and other large species from the superficial deposits. Lastly, Cuvier claimed, came the bones of mammals of species still known alive, but they were only from the geologically recent alluvial deposits. Having sketched this history of the vertebrates, however, Cuvier pointed out that its zoological dimension—that is, the product of his own research—was more firmly established than the geological, for which he had had to rely on collectors who had often neglected to record the location of their finds with sufficient precision.

At this point Cuvier launched into a major counterattack against the criticism of his work that had been made while his papers were being published in the *Annales,* on the grounds that the species he claimed as extinct could well be the ancestors of living species, transformed in the course of time, and therefore not the victims of any "catastrophe" (sec. 33). Although transformist theorizing had been fashionable for many years among some Parisian naturalists, Cuvier's wording leaves little doubt that his colleague Lamarck was his chief target. His defense of the integrity and stability of animal species led him into an important discussion of the range of intraspecific variation in living forms, and particularly of the effects of climate and domestication. He concluded that at least under natural conditions variation was strictly limited, far too much so to have ever allowed the transformation of one species into another.[4]

As for the temporal dimension of the argument, Cuvier invoked the evidence of the mummified animals from ancient Egypt as the best test case available. Since there was no difference at all between mummified and living individuals of the same species, there was no good reason for claiming that the lapse of much longer periods of time would turn one species into another. For Cuvier this involved a fundamental point of scientific method: handwaving about the inconceivable magnitude of geological time was vacuous, if the effects claimed could not be demonstrated on a smaller scale within a shorter time.[5] The strategic impor-

4. In effect this discussion set the agenda for research on the species problem, not only for naturalists of Lyell's generation, but right up to the time of Darwin.

5. Although it is not explicit, Cuvier may well have had in mind the analogy with astronomy that he had used in the opening of the "Discourse" (sec. 1). Just as the mathematical astronomers could extrapolate from accurate observation of the movement of a planet over a few years, to infer an orbit lasting many centuries, so (Cuvier argued) naturalists should only extrapolate into the vastness of geological time what they could demonstrate with precision for a shorter period. Lamarck, on the other hand, claimed that the absence of any perceptible organic change over the past few thousand years was proof that the process operated far more slowly: in effect it was like the absence of parallax

tance of Cuvier's earlier paper on the one apparent exception to the rule about the mummified animals became apparent here; it was not only for reasons of logical order that he had his paper on the ibis bound into the first volume of *Ossemens fossiles,* immediately following the "Discourse" (see figs. 23, 24).[6]

Among the living species distinct from any fossils, the human species was of course of paramount interest. Cuvier therefore turned next to the question of the existence of human fossils (sec. 34). After a careful review of the bones that had been claimed as human, he concluded that none was authentic, and he inferred that the continents on which the bones of extinct mammals are now found were not inhabited at that time by any human beings. The period before the "catastrophe" that had made those mammals extinct was not, however, wholly prehuman: Cuvier conceded that human beings might have lived elsewhere at that time, but if so their fossil remains must lie undiscovered, perhaps on the floors of present oceans.

In any case, what mattered here to Cuvier was that the fossil bones he had studied could not be used to support claims for an immense antiquity for any existing human culture. On the contrary, everything pointed to the geologically recent date at which the present continents had taken roughly their present form (sec. 35). Here Cuvier borrowed explicitly from the work of Deluc and Dolomieu, to show that geological processes now at work on the continents could not have been acting for more than a geologically short time.[7]

This physical evidence was next supported by the textual evidence of human history (sec. 36). As in his argument against transformism, Cuvier here embroiled himself in a major current debate, this time the one about human chronology. In parallel with his debunking survey of alleged human fossils, here he rejected all claims to see in human records any evidence that would carry literate human societies back more than a few thousand years. To make this point, he surveyed the whole range of human records known in his time.

---

in observations of the distant stars. The methodological difference between the two savants was perhaps irreconcilable, but Cuvier's stance certainly deserves as much respect as Lamarck's.

6. It was logical to put this paper in the first volume of *Ossemens fossiles,* because all the other volumes dealt with truly *fossil* bones (and those of their living "analogues").

7. Much of this evidence remained valid when, later in the century, geologists argued that the Pleistocene glaciations had ended only a few thousand years ago. On the glacial theory, the change from a glacial (or at least periglacial) climate to a temperate one, in the areas of Europe that Cuvier and his contemporaries knew best, had indeed been so recent that the processes now acting had had only a geologically short time in which to produce their observed effects.

The records of ancient Jewish history came first, but primarily because they were the oldest to be intelligible. They were handled in exactly the same way as those of other cultures; indeed, Genesis was treated just as a kind of proxy for Egyptian records, which were considered to be still older but had not yet been deciphered. That Cuvier was no covert biblical literalist is amply indicated by the fact that he cited a leading exponent of the new German biblical criticism as his authority for dating the text of the Pentateuch, including Genesis. That is not to say that Cuvier rejected the biblical story of the Flood as merely fabulous: on the contrary, his wording leaves no doubt that he regarded it as a faint *textual* record of the very "catastrophe" of which his geological research had uncovered the *physical* traces. But in his view the textual record of the oldest part of the Old Testament was in no way superior—as a *scientific* source—to those of other ancient cultures: *all* were "incoherent traditional stories" that had to be squeezed hard to yield any historical value at all.

Cuvier's review of ancient cultures took him far beyond the Mediterranean, to India and even China: once again, his work for the French edition of *Asiatick researches* clearly stood him in good stead. He concluded that the textual records of *all* known cultures pointed to a relatively recent origin for human civilization in its present form, compatible with the physical evidence for a geologically recent "grand revolution in nature." Two other objections to such a conclusion were dismissed briefly (secs. 37, 38): neither an inscribed zodiac brought from an Egyptian temple nor the spoil heaps of an ancient mine in Italy could in Cuvier's view support the very high antiquity that had recently been claimed for them.

In his general conclusion (sec. 39), Cuvier therefore asserted that a "great and sudden revolution," only a few thousand years ago, had in effect reversed the positions of continents and oceans. His earlier notion of a brief incursion of the sea, over continents that otherwise remained relatively unchanged (text 11), had now been replaced by a reversion to the much more radical theorizing of Deluc and Dolomieu. The present continents had been ocean floors before the catastrophic event, and had afterward been populated by men and animals that had survived from the previous continents, areas that were now beneath the sea. Yet the fossils found on the present continents indicated that at a still earlier time, before that marine period, those areas had been dry land; and Cuvier alluded to his and Brongniart's work on the Paris region (text 15) to argue that there seemed to have been several such alternations. As usual, he declined to make any suggestion about the possible *cause* of these major physical changes: he pointed out correctly that in a *historical* science—such as geology had by now become—the reality or historicity of the events themselves had first to be established, from an analysis of their effects, before attempting to find causal explanations for them.[8]

If Cuvier's conception of the nature of the "catastrophes" remained vague, incoherent, and even inconsistent at the end of the "Discourse," his prescription for future research was clear and unambiguous (sec. 40). In the closing pages he again set out the research *agenda* he had presented five years previously at the Institut (text 13) and elaborated earlier in the "Discourse" (secs. 22–25). It focused once more on the need for much closer attention to the relation between fossils and the rock formations in which they are found, and particularly on the relatively recent formations that linked the ancient history of the earth to the present. Here the formations of the Paris region served explicitly as a model: the full version of his and Brongniart's paper on them (text 15) was bound with the "Discourse" in the first volume of *Ossemens fossiles*. These formations were all younger than the Chalk, which until their joint research had generally been regarded as the most recent "regular" formation. Cuvier suggested that their paper should be extended with studies of other relatively young formations, such as those in the foothills of the Apennines; his own brief inspection of their fossils—on his trip to Italy in the service of educational administration—had convinced him that they would provide the final link between the most ancient formations and the current processes at work on earth.[9]

Cuvier ended the "Discourse," as he had begun it, with the analogy with human history. For geologists to focus attention—as they did—on the old Primary rocks, which contained no fossils, was as if scholars were to lose interest in French history at just the point when the arrival of the literate Romans first provided documentation for that history. Geologists had lost themselves in a morass of conjectures about the origin of the

8. A more modern analogy would be the insistence of many earlier twentieth-century geologists and paleontologists (at least outside the United States) that there was strong evidence for the reality of some kind of crustal mobilism ("continental drift"), which deserved to be taken seriously even in the absence of any proposed mechanism that would satisfy the geophysicists; the mechanism (of "plate tectonics") duly came later.

9. They were already known to contain many more shells of extant species than the Paris formations. Cuvier's suggestion was duly taken up by Giovanni Battista Brocchi (1772–1826), the director of the new natural history museum in Milan, where the best collection of the shells was kept (possibly Cuvier had got the idea from him in the first place): Brocchi's great monograph on the fossil shells of the sub-Apennine hills was published two years later (*Conchiologia fossile subappenina*, 1814). Years later still, Lyell too took the hint and made the same formations, and similar ones in Sicily, the centerpiece of his interpretation of *all* the Tertiary strata of Europe.

earth, just as speculative historians had wasted effort on conjectures about the origins of society. Cuvier's research program proposed a more fruitful way forward: to penetrate backward from the present into the relatively recent past, before trying to tackle the more problematic older formations. Above all, however, Cuvier's work fleshed out the metaphor that had become almost a cliché: the naturalist was to use fossils as the historian used documents, to piece together an authentic *history* of the earth, and of life at its surface. To "burst the limits of time," as he had put it at the start, was to write a *prehuman* history.

TEXT 19

———

Preliminary Discourse

[1 INTRODUCTION]

I shall try to travel a road on which only a few steps have so far been ventured, and to make known a kind of monument that is almost always neglected, although it is indispensable for the history of the globe [fig. 22].

As a new species of antiquarian, I have had to learn to decipher and restore these monuments, and to recognize and reassemble in their original order the scattered and mutilated fragments of which they are composed; to reconstruct the ancient beings to which these fragments belonged; to reproduce them in their proportions and characters; and finally to compare them to those that live today at the earth's surface. This is an almost unknown art; and it presupposes a science hardly touched on *[effleurée]* hitherto, namely that of the laws that govern the coexistence of the forms of the different parts of organisms. I therefore had to prepare myself for this research, by much more lengthy studies of existing animals. An almost general review of the present creation *[création actuelle]* could alone give demonstrative character to my results on this ancient creation *[création ancienne]*. At the same time, however, this review has given me a great collection of rules and relationships [10] no less demonstrated; so that the whole

---

10. [Cuvier first wrote "a complete system *[système entier]* of rules and relationships"; the change indicated a sense of caution, and perhaps a desire to avoid the term "system," which he had criticized so relentlessly in geology.]

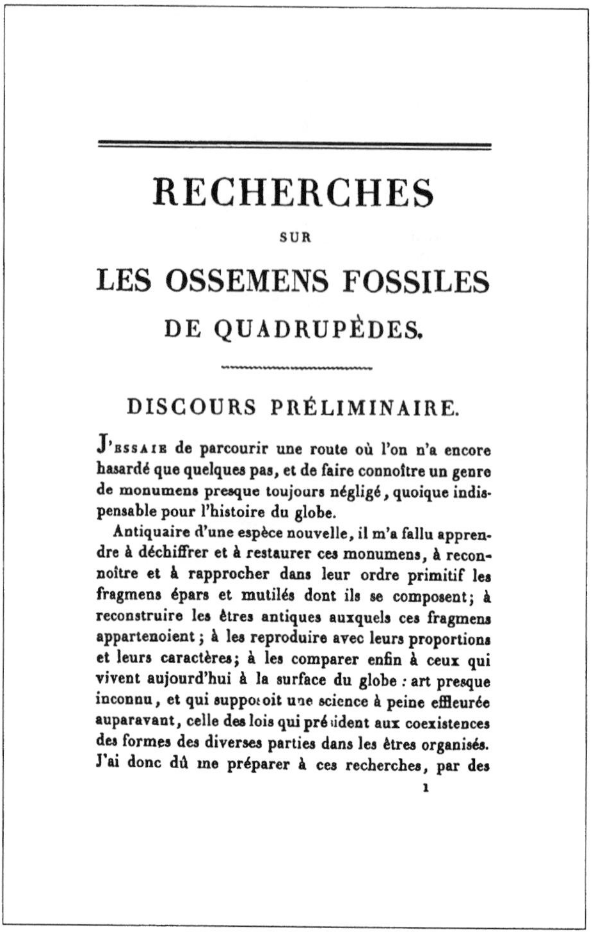

RECHERCHES

SUR

LES OSSEMENS FOSSILES

DE QUADRUPÈDES.

DISCOURS PRÉLIMINAIRE.

J'ESSAIE de parcourir une route où l'on n'a encore hasardé que quelques pas, et de faire connoître un genre de monumens presque toujours négligé, quoique indispensable pour l'histoire du globe.

Antiquaire d'une espèce nouvelle, il m'a fallu apprendre à déchiffrer et à restaurer ces monumens, à reconnoître et à rapprocher dans leur ordre primitif les fragmens épars et mutilés dont ils se composent; à reconstruire les êtres antiques auxquels ces fragmens appartenoient; à les reproduire avec leurs proportions et leurs caractères; à les comparer enfin à ceux qui vivent aujourd'hui à la surface du globe : art presque inconnu, et qui supposoit une science à peine effleurée auparavant, celle des lois qui président aux coexistences des formes des diverses parties dans les êtres organisés. J'ai donc dû me préparer à ces recherches, par des

1

FIGURE 22   The opening page of Cuvier's "Discours préliminaire" (1812), with his famous identification of himself as an "antiquaire d'une espèce nouvelle" (new species of antiquarian). The handsome format indicated that the work bore all the authority and prestige of the Muséum and the Institut; the typography was the same as that used in the *Annales du Muséum,* from which the papers in the rest of *Ossemens fossiles* had been reprinted.

animal kingdom is found to be subject to new laws, on the occasion of this essay on a small part of the theory of the earth.[11]

The importance of these truths, which developed in proportion as I advanced in my work, has helped to sustain my efforts, no less than the

11. This will be seen in my large *Comparative anatomy* [*Règne animal,* 1817], on which I have been working for more than twenty-five years, and which I intend to begin publishing shortly.

novelty of my main results. Would that it could have a similar effect on the steadfastness of the reader, and persuade him to follow, without too much weariness, the arduous paths on which I am obliged to take him!

Moreover, the ancient history of the earth, the ultimate goal toward which all this research is leading, is in itself one of the most fascinating subjects on which the attention of enlightened men [hommes éclairés][12] can be fixed. If they take an interest in following, in the infancy of our [own] species, the almost erased traces of so many extinct nations [nations éteints], they will doubtless find it also in gathering, in the darkness of the earth's infancy, the traces of revolutions previous to the existence of every nation.

We admire the power by which the human spirit has measured the movements of the globes, which nature seemed to have concealed forever from our view; genius and science have burst the limits of space, and some observations developed by reason have unveiled the mechanism of the world. Would there not also be some glory for man to know how to burst the limits of time, and, by some observations, to recover the history of the world, and the succession of events that preceded the birth of the human species? The astronomers have without doubt progressed more rapidly than the naturalists; and the stage at which the theory of the earth[13] currently finds itself is a little like that in which some philosophers believed the heavens to be made of dressed stone, and the moon to be as large as the Peloponnese.[14] But after the Anaxagorases came the Copernicuses and the Keplers, who cleared the way to Newton; so why should not natural history also have its Newton one day?[15]

What I present here comprises only a very small part of the facts that this ancient history should embrace. But these facts are important: several of them are decisive, and I hope that the rigorous way in which I have

12. [Cuvier first wrote just "human attention" (attention humain), perhaps before accepting that such a sense of curiosity was not, unfortunately, a universal human characteristic.]

13. [Cuvier first wrote "geology" rather than "theory of the earth"; in reading the "Discourse" it is important to remember that the word "geology" was still widely used as a synonym for the ambitious theoretical project about which Cuvier had earlier been so scornful.]

14. [The Greek philosopher Anaxagoras had claimed that the sun (not the moon) was no larger than the Peloponnese (the southern part of modern Greece), and was accused of impiety for his pains.]

15. [This final clause was an addition to the original manuscript text. In context it would seem to express an ambition to find geological laws for the universe of time, comparable to Newton's laws of gravitation for the universe of space. However, since the aspiration was phrased in terms of "natural history," Cuvier may also have had in mind Immanuel Kant's well-known rejection of any hope that a Newton-like figure would ever formulate natural laws for biology. It is tempting to infer that Cuvier, who did not count modesty among his virtues, saw himself as prime candidate for this honor; but his consistent caution about going beyond purely phenomenological laws, not least in his geology, suggests that he saw himself rather as the equivalent of a Kepler.]

proceeded to identify them will allow them to be considered as points that are definitely fixed and from which it will no longer be permissible to depart. If this hope is justified only in relation to a few of them, I shall consider myself adequately rewarded for my trouble.

[2] OUTLINE [OF THE ARGUMENT]

In this preliminary discourse I shall recount the set of conclusions that it seems to me that the theory of the earth has currently reached. I shall show the connections linking these results to the history of the fossil bones of land animals, and the reasons that give particular importance to that history. I shall expound the principles underlying the art of identifying these bones, or, in other words, of recognizing a genus and distinguishing a species from a single fragment of bone: on the certitude of this art rests that of the whole work. I shall expound in rapid manner the results of the researches that the work comprises; the new species, and genera formerly unknown, that these researches have led me to discover; the different kinds of formation that contain them; and, as the difference between these species and those of today does not exceed certain limits, I shall show that these limits go far beyond those that today distinguish the varieties of a single species. Thus I shall make known how far this variation can extend, whether by the influence of time, or by that of climate, or finally by that of civilization.

In this way I shall put myself in a position to conclude that great events were necessary to bring about the major differences I have recognized. I shall thus expound the particular ways that my work makes it necessary to modify hitherto received opinions on the primitive history of the globe. Finally I shall examine how far the civil and religious history of human societies [peuples] accords with the results of observation on the physical history of the earth, and with the probabilities that these observations provide concerning the epoch at which human societies were able to find fixed abodes and fields amenable to cultivation, and at which in consequence they were able to take a durable form.

[3] THE EARTH AT FIRST GLANCE

When a traveler crosses fertile plains, where the regular course of tranquil rivers sustains abundant vegetation, and where the land—crowded with numerous people and ornate with flourishing villages, rich cities, and superb monuments—is never disturbed unless by the ravages of war or by the oppression of powerful men, he is not tempted to believe that nature has also had its civil wars, and that the surface of the globe has been upset by successive revolutions and various catastrophes. But these ideas change as soon as he seeks to excavate this ground that today is so peaceful, or to climb onto the hills that border the plain. His ideas enlarge, as it were, with his viewpoint. They begin to encompass the extent and magnitude of these ancient events, as soon as he climbs the higher chains of which the [foot]hills cover the flanks, or as he penetrates into their interior, following downward the beds of the torrents.

[4] INITIAL EVIDENCE OF REVOLUTIONS

The lowest, most undisturbed [unis] terrains, when penetrated to very great depths, show only horizontal beds of various materials, almost all containing the innumerable [organic] products of the sea. Comparable beds and similar products compose the hills up to great heights; sometimes the shells are so numerous that by themselves they make up the whole mass of the ground. Almost everywhere they are so well preserved that the smallest of them retain their most delicate parts, their most subtle ridges and finest points. They are raised to greater elevations than the level of any seas, and where no sea could today be carried by any existing causes. They are contained not only within loose sands; the hardest stones often encrust them and penetrate every part of them. All parts of the world, all hemispheres, all continents, all islands of any size: all show the same phenomenon.

One is thus soon disposed to believe, not only that the sea invaded all our plains, but also that it stayed there long and peacefully, so as to form deposits so thick and extensive, in part so solid, and containing such well-preserved remains. The time is past when ignorance could maintain that these remains of organisms were simply sports of nature, products conceived within the bowels of the earth by its creative forces. A careful comparison of their forms and their tissue, even often of their composition, shows not the slightest difference between these shells and those that the [present] sea sustains. Thus they lived in the sea; they were deposited by the sea; the sea existed in the places where it left them; and the basins of the sea have at least undergone a change in either extent or situation. That is what already results from an initial excavation, and from the most superficial observation.

The traces of revolutions become more imposing as one rises a little higher, and approaches nearer the foot of the large [mountain] chains. There are still indeed shelly beds [bancs], and one even sees thicker and more solid ones there; and the shells are just as numerous and well preserved. But they are no longer the same species; the beds that contain them are no longer so generally horizontal, but are tilted obliquely and sometimes almost vertically. Whereas it was necessary in the plains and low hills to dig deeply in order to discover the succession of beds, here they can be seen on the flanks, by following the valleys produced by their fractures [déchiremens]. At the feet of their escarpments, immense masses of their debris form rounded hills, which increase in height with every thaw and every storm.

These tilted beds, which form the crests of the Secondary mountains, are not placed above the horizontal beds of the hills that form their feet; on the contrary, they plunge below them. Those foothills lean against their slopes. When one penetrates the horizontal beds in the vicinity of the oblique ones, one finds the latter at depth; sometimes, when the oblique beds are not too elevated, their summit is even capped with horizontal beds. The oblique beds are thus more ancient than the horizontal; and since they must have been formed horizontally, they have been tilted up; they were [raised] before the others leaned against them.

Thus the sea, before it formed the horizontal beds, had formed others, which by some cause or other have been broken, tilted, and disturbed in a thousand ways. There was thus also at least one change in the bosom of the sea that preceded ours; it too has undergone at least one catastrophe. And,[16] since many of these oblique beds that it had formed first rise above the horizontal beds that succeeded them and surround them, the catastrophe that made these beds oblique also thrust them above the level of the sea, making them islands or at least reefs and inequalities, whether they were elevated at one end, or the subsidence of the other end had lowered the waters. This second result is no less clear and demonstrable than the first, for anyone who takes the trouble to study the monuments that support it.

[5] PROOFS THAT THESE REVOLUTIONS HAVE BEEN NUMEROUS

If the various beds and the organic products they contain are compared with one another in greater detail, however, one soon perceives still more numerous differences, which indicate still more numerous changes of state. The sea did not continuously deposit rocks similar to each other; there is a regular succession in the nature of its deposits. The more ancient the beds, the more each of them is uniform over a wide area; the newer the beds, the more limited they are and the more subject they are to variation over small distances. Thus the great catastrophes that produced revolutions in the

16. [The remainder of this paragraph was an addition to Cuvier's original manuscript text. Although rather obscure, it does show him grappling with the dynamics of the movements revealed by the tilted rocks, and at least considering the possibility that there had been forceful uplift of the crust as well as passive collapse.]

basin of the seas were preceded, accompanied, and followed by changes in the nature of the liquid and in the materials that it held in solution; and when the surface of the sea had been divided up by islands, by projecting chains, there were different changes in each separate basin.

During such changes in the general liquid, it was very difficult for the same animals to continue to live in it. And they did not do so. Their species and even their genera change with the beds. Although there might be some returns of species at small distances [i.e. from one bed to another], it is true to say that in general the shells in the ancient beds have forms that are specific to them, and that they disappear gradually and are no longer found in the recent beds. Still less are they found in present seas, where the analogues of their species are never discovered, where even many of their genera are not found. The shells of the recent beds, by contrast, resemble in their genera those that are alive in the seas; and in the last and least consolidated *[les plus meubles]* of these beds, there are some species that the best-trained eye cannot distinguish from those that the ocean [now] sustains.

Thus there has been a succession of variations in animal nature, corresponding to those in the chemical nature of the liquid; and when the sea left our continents for the last time, its inhabitants did not differ much from those that it still sustains today.

Finally, if these remains of organisms are examined with still greater care, one discovers in the middle of the marine beds—even in the most ancient— some beds that are filled with animal or plant productions of the land or of freshwater; and it is among the most recent beds, that is, the most superficial, that terrestrial animals are buried under masses of marine products. Thus the various catastrophes of our planet have not only caused the different parts of our continents to emerge by degrees from beneath the waves; it has also happened several times that areas *[terrains]* made into dry land have been covered again by water, either by having subsided or because the water has simply risen over them. And the particular ground that the sea left during its last retreat had already been dried out once before, and had sustained quadrupeds, birds, plants, and all kinds of terrestrial productions; it had thus been invaded by the sea that has since left it.[17]

The changes that have happened in the productions of the shelly beds have therefore not depended solely on a gradual and general retreat of the waters, but on various successive advances *[irruptions]* and retreats; the final result, however, has been a universal lowering of sea level.

17. [The reference in this paragraph is of course to the formations of the Paris Basin that Cuvier and Brongniart had studied in detail, and that had provided Cuvier with some of his most important fossil vertebrates (text 15).]

[6] PROOFS THAT THESE REVOLUTIONS HAVE BEEN SUDDEN

These repeated advances and retreats have not been slow at all, nor achieved by degrees; most of the catastrophes that led to them have been sudden. This is above all easy to prove for the last of all of them, the traces of which are most discoverable. Moreover, in the northern countries it has left the carcasses of large quadrupeds that the ice seized, which have been preserved to this day with their skin, hair, and flesh. If they had not been frozen as soon as they were killed, putrefaction would have decomposed them. Now this eternal frost could only have taken hold of the places where these animals lived by the same cause that destroyed them: [18] the cause was thus as sudden as its effect. The tearing and upheaval of beds that happened in the earlier catastrophes show that they were as sudden and violent as the last one; and the masses of debris and rolled stones, which in many areas are found between the solid beds, attest to the force of the movements that these upheavals generated in the body of water.

Thus life on earth has often been disturbed by terrible events: calamities which initially perhaps shook the entire crust of the earth to a great depth, but which have since become steadily less deep and less general. Living organisms without number have been the victims of these catastrophes. Some were destroyed by deluges, others were left dry when the seabed was suddenly raised; their races are even finished forever, and all they leave in the world is some debris that is hardly recognizable to the naturalist.

Such are the consequences to which the objects we encounter at every step necessarily lead, and that we can verify at every moment in almost every land. These great and terrible events are clearly imprinted everywhere, for the eye that knows how to read history in their monuments. But what is still more astonishing, and no less certain, is that life has not always existed on the globe, and that it is easy for the observer to recognize the point at which it began to deposit its products.

[7] PROOFS THAT THERE WERE REVOLUTIONS BEFORE ORGANISMS EXISTED

Let us climb farther, toward the main ridges, toward the high summits of the large [mountain] chains. Soon these remains of marine animals, these

---

18. The two most remarkable phenomena of this kind, which must forever banish all idea of a slow and gradual revolution, are the rhinoceros discovered [by Pallas] in 1771 in the banks of the Vilhoui, and the elephant recently found by Mr. Adams near the mouth of the Lena [see chapter 9]. The latter still retained its flesh and skin, on which the hair was of two kinds; one short, fine, and crisp, resembling wool, and the other like long bristles. The flesh was still in such a good state of preservation that it was eaten by dogs.

innumerable shells, become rarer and disappear completely; we reach beds of another kind, which contain no vestiges of any organisms. Nevertheless, these show by their crystalline character, and indeed by their stratification, that they too were formed in a liquid; by their tilted position and their escarpments, that they too have been upheaved; by the way they plunge obliquely beneath the shelly beds, that they were formed before the latter; and finally, by the height to which their bristling and naked peaks rise above all the shelly beds, that their summits have not been covered by the sea since their elevation made them emerge.

Such are the famous Primitive or Primordial mountains that cross our continents in various directions, rising above the clouds, separating the river basins, holding in their perpetual snow the reservoirs that feed the springs, and forming as it were the skeleton or gross structure of the earth. From a great distance the eye can see—in the jaggedness that breaks up their crest, in the sharp peaks that bristle on it—signs of the violent way in which they were elevated: quite different from those rounded mountains, and those hills with broad flat surfaces, whose recent mass has always remained in the situation in which it was deposited quietly by the last seas.

These signs become clearer in proportion as one approaches. The valleys no longer have the gently sloping flanks, with projecting shoulders alternating with one another, that seem to indicate the beds of some ancient streams. They become wider or narrower without any regularity; their waters sometimes expand into lakes, sometimes fall in torrents; sometimes the rocks, suddenly drawing closer together, form transverse dikes, over which those same waters fall in cataracts. At the peak their shattered beds show a sheer edge on one side and large oblique areas of their surface on the other; their heights do not correspond at all; but those that on one side form the crest of the escarpment are often sunken on the other in such a way as to disappear.

In the midst of all this disorder, however, some naturalists have believed they could see that a certain order still reigns, and that these immense formations [bancs]—broken and overthrown though they are—observe among them a sequence that is more or less the same in all [mountain] chains. They [the naturalists] say that granite, which rises above all, also plunges beneath all the others; it is the most ancient of the rocks that it has been given to us to see in the place that nature assigned them. The central ridges of most chains are composed of it; the stratified rocks lean against its flanks and form lateral ridges; schists, sandstones, and talcic rocks join their beds to it; and finally, leaning against the schists and forming the outer ridges, are granular marbles and other limestones without shells, which are the last work by which this lifeless sea seems to have been prepared for the production of its shelly beds.[19]

Even in regions far from the major [mountain] chains, wherever one can pierce through the recent beds and penetrate some way into the thickness of the crust [enveloppe] of the globe, more or less the same order of stratification is found. Granular marbles never cover shelly beds, massive granites never rest on granular marbles (except in a small number of places where there seems to be granite of several epochs): in a word, this whole arrangement seems to be general, and must therefore be due to general causes, which have each time exercised their influence from one end of the earth to the other. Thus it is undeniable that for a long time the waters covered the formations [masses] that today form our highest mountains; and that for a long time these waters sustained no living organisms whatever.

It is not only since the birth of life that changes of nature and numerous revolutions have occurred: the formations formed previously varied as much as those formed since; they have also suffered violent changes in their position, and some of these changes took place at the time when these rocks existed alone and had not been covered at all by the shelly rocks. There is proof of this in the upheavals, ruptures, and fissures that are seen in their beds, in even greater number and more marked than in those of subsequent formations [terrains].

These same Primitive rocks, however, have sustained still more revolutions since the production of the Secondary formations, and have perhaps occasioned – or at least shared in – those that the latter have suffered. In fact, considerable portions of the Primitive formations are exposed, although in a lower situation than many of the Secondary formations: how would the latter not have covered those parts, if they were not exposed since the Secondary rocks were formed? In certain areas numerous large blocks of Primitive rocks are widespread on the surface of the Secondary formations, separated by deep valleys from the peaks or ridges from which these blocks could have come: either eruptions must have thrown them there, or the valleys that would have stopped their course did not exist at the epoch of their transport.[20]

There is a collection of facts, then, and a series of epochs anterior to the present time, the order of which can be verified without uncertainty, although the duration of the intervals between them cannot be defined with

19. See Pallas, in his memoir "On the formation of mountains" ["Observations sur la formation des montagnes," 1778].

20. The journeys of Saussure [1779–96] and Deluc [1778] provide a mass of examples of this kind. [The reference is to the "erratic blocks" that were later—after Cuvier's death—attributed to transport by vanished glaciers and ice sheets.]

precision. They are so many points that will serve to give measure [règle] and direction to this ancient chronology [antique chronologie].

[8] EXAMINATION OF THE CAUSES THAT TODAY
STILL OPERATE AT THE EARTH'S SURFACE

Let us now examine what takes place on earth today; let us analyze the causes that still operate at its surface and determine the possible extent of their effects. This part of the [natural] history of the earth is all the more important, since it has long been thought possible to explain earlier revolutions by these present causes, just as past events in political history are easily explained when one knows well the passions and intrigues of our times. But we shall see that unhappily this is not so in physical history.[21] The thread of operations is broken; nature has changed course, and none of the agents she employs today would have been sufficient to produce her former works.

There now exist four active causes that together alter the surface of our continents: the rains and thaws that wear down steep mountains and wash the debris to their feet; the running waters that carry this debris and deposit it in places where the current slackens; the sea, which undermines the base of elevated coastlines to form cliffs, and which throws up sand dunes on flat coastlines; and finally the volcanos that pierce the solid strata and build up or spread masses of ejected material there.

[9] LANDSLIDES [ÉBOULEMENS]

Everywhere that the edges of broken strata are exposed on sheer [cliff] faces, fragments of their material fall to their feet every spring, or even in every storm; they are rounded by rolling over each other, and the whole mass adopts a slope determined by the laws of cohesion. Thus there is formed at the foot of the escarpment a more or less elevated scree [croupe], depending on the abundance of the falls of debris. These screes form the flanks of the valleys in all high mountains, and are covered with rich vegetation when landslides from above begin to become less frequent. But their lack of solidity makes them liable to further slides when they are undermined by streams. It is then that towns, and rich and populous districts, find themselves buried under a mountain's landslide; the course of rivers is intercepted; and lakes are formed in previously fertile and pleasant places. But happily these big landslides are rare, and the main effect of these hills of debris is to provide materials for the ravages of torrents.

---

21. [Cuvier first used a striking metaphor when drafting this phrase: "the history of nature is private [privé]."]

[10] ALLUVIA

The waters that fall on the ridges and peaks of mountains, or the vapors that condense there and the snows that melt there, run down their slopes in an infinity of trickles, removing some fragments and marking their passage with shallow furrows. Soon these trickles unite into the deeper gullies with which the surfaces of mountains are furrowed; they flow out into the deep valleys that cut into their feet, and thus form the streams and rivers that carry to the sea the waters that the sea had given to the atmosphere. When the snow melts, or when there is a storm, the volume of these mountain waters is suddenly increased, and they rush down with a speed proportional to the slope. They strike violently against the feet of the screes that cover the flanks of all the high valleys; they carry away the already rounded fragments of which the screes are composed; and they blunt and polish them further by friction. But as they reach the broader valleys where the slope is reduced, or in wider basins where they are able to spread out, they drop on the banks the heaviest of the stones that they were rolling along; the smaller debris is deposited lower down; and hardly anything but the most minute fragments, or the most imperceptible silts, reach the main channel of the river. Often, even before forming the large lower river, the water has to cross a vast and deep lake, where its silt is deposited, so that it reemerges limpid. The lower rivers, and all the streams that rise in the lower mountains or hills, also produce, in the land they traverse, effects more or less analogous to those of the torrents of high mountains. When they are swollen by heavy rains, they attack the feet of the earthy or sandy hills that they meet in their course; they carry the debris on to the low ground that they flood, which is raised somewhat by each inundation. Finally, when the rivers reach a large lake or the sea, and the flow that carried the fragments of silt ceases altogether, these fragments are deposited on the sides of the river mouth. They end by forming land there, which extends the coastline; and if this coastline is such that the sea on its side throws up sand and contributes to this growth, provinces or entire kingdoms are created. Ordinarily these are the most fertile, and will soon be the richest in the world, if governments allow industry to be practiced in peace.[22]

[11] DUNES

The effects that the sea produces without the assistance of rivers are much less happy. When the coast is low and the sea bottom sandy, the waves push

22. [Probably an allusion to the fertile provinces of the Netherlands, and the rich Po delta in Lombardy in northern Italy, both of which Cuvier had seen at first hand in 1809–11, in the course of his travels for Napoleon's government.]

this sand toward the shore; at each low tide it dries out a little, and the wind that almost always blows from the sea throws it onto the beach. Thus dunes are formed, those sandy mounds which—if human industry does not manage to fix them with appropriate plants—move slowly but inexorably toward the interior of the land, covering fields and dwellings; because the same wind that lifts the sand from the shore onto the dune throws it from the crest of the dune onto the slope away from the sea.

[12] CLIFFS

When the coast is elevated, the sea cannot cast anything up on it, and on the contrary exercises a destructive action. Its waves gnaw at the base and cut the whole height into a cliff; because the higher parts, being deprived of support, fall into the water. There they are swirled by the waves until the softest and loosest fragments disappear. The harder portions, through being rolled in opposite directions by the waves, form those rounded pebbles that the sandy shore in the end accumulates sufficiently to act as a rampart at the foot of the cliff.

Such is the action of water on dry land; and it can be seen that it consists of almost nothing but levelings—and of levelings that are not indefinite. The debris of the major ranges are swept into the small valleys; their particles, and those of the hills and plains, are carried down to the sea; the alluvia extend the coasts at the expense of the high ground. These are the limited effects, which vegetation generally brings to an end, and which anyway require the preexistence of the mountains, valleys, and plains, and which in consequence cannot have given birth to these inequalities of the globe. Dunes are an even more limited phenomenon, both in height and in horizontal extent; they have no relation at all to the enormous formations [*masses*] for which geology seeks an origin.

As for the action that the seas have in their own realm, although it cannot be ascertained so well, it is nonetheless possible up to a certain point to determine its limits.

[13] DEPOSITS UNDER WATER

Where streams discharge into lakes, [water] meadows, marshes, and seaports, and above all when they flow down from steep hillsides nearby, they deposit masses of silt at the bottom, which would eventually fill them up if care were not taken to clean them out. The sea likewise deposits mud and sediment in ports, coves, and every place where the water is calmer. Currents amass in their midst, or deposit on their sides, the sand that they tear from the sea bottom, and build up sandbanks and shallows from it.

[14] PRECIPITATES [STALACTITES][23]

Certain waters, having dissolved calcareous substances by means of the superabundant carbonic acid they contain, leave them to crystallize when that acid is able to evaporate, forming stalactites and other concretions. There are some beds that are confusedly crystallized in freshwater, extensive enough to be comparable to some of those that the former sea left behind.

[15] CORALS [LITOPHYTES][24]

In the tropical zone, where corals are numerous in species and propagate with great vigor, their stony stems are interlaced with rocks to form reefs; rising to the water level, they close the entrances to ports, setting terrible traps for navigators. The sea, throwing sand and silt on top of these reefs, sometimes elevates the surface above its own level, forming islands that a rich vegetation soon comes to enliven.

[16] INCRUSTATIONS

It is also possible that in some areas shelled animals leave behind their stony remains after death; and that, bound by more or less solid mud or by other means, they form extensive deposits or kinds of shell banks. But we have no proof that the sea today can cement these shells with a paste as compact as that which we see enveloping our beds of marble [marbres] and sandstone, or even the Coarse Limestone [calcaire grossier].[25] Still less do we find that it precipitates any part of the [even] more solid and more siliceous beds that preceded the formation of the shelly beds.[26] In short, all these causes taken together would not produce a single bed, or elevate the smallest hillock, or change the level of the sea to any appreciable degree.

It has indeed been claimed that the sea is undergoing a general diminu-

---

23. [As the paragraph below shows, the word "stalactite" was used at this time (in both languages) to mean chemically precipitated rocks, of which the icicle-like structures in caves were just one variety.]

24. [The standard term for stony corals (again in both languages) still preserved the earlier conjecture that they were stony *plants*, although their animal character was by this time well recognized by naturalists.]

25. [The word "marble" *(marbre)* was used here in the everyday sense, to mean any limestone hard and dense enough to take a polish (rather than in the modern technical sense of a limestone that has been recrystallized by metamorphism); the *calcaire grossier* of the Paris Basin (see text 15) was relatively soft and rubbly.]

26. [That is, the Primary and Secondary formations respectively. At this time there was no adequate understanding of the chemical changes responsible for compacting sediments into solid rocks. Therefore the hardness and solidity of most sedimentary rocks accentuated the apparent contrast between the older formations and what the present seas seemed to be capable of producing.]

tion, and that this has been observed in some places on the shores of the Baltic.[27] But whatever may be the cause of this appearance, it is certain that nothing similar has been observed on our coasts, and that there is no general lowering whatever of the waters. The oldest seaports still have their quays and all [other] constructions at the same height above sea level as at the time when they were built.

Some general movements of the sea from east to west, or in other directions, have indeed been alleged; but it has nowhere been possible to estimate such effects with any precision.[28]

[17] VOLCANOS

The action of volcanos is even more limited and local than all of those we have just spoken about. Although we have no idea of the means by which nature maintains these violent furnaces [foyers] at such great depths, from their effects we can clearly judge the changes they have been able to produce at the surface of the globe. When a volcano breaks out, after some shocks and earthquakes, it makes an opening for itself. Rocks and ashes are thrown afar. Lavas are spewed out; the more fluid parts flow out in long currents, while the less fluid parts stop at the rim of the opening and raise its edge, forming there a cone topped with a crater. Thus volcanos, having modified materials formerly buried at depth, accumulate them at the surface; they form mountains; in the past they covered some parts of our continents; and they have given sudden birth to islands in the middle of the sea.[29] But these mountains and islands were always composed of lavas; all their materials had been subjected to the action of fire. Volcanos thus neither elevate nor knock down the beds that traverse their orifice [soupirail], and they have not contributed at all to the elevation of high nonvolcanic mountains.

27. [A reference to the famous claim by the Swedish savant Anders Celsius (1701–44), which was based on accurate historical records of changes in the heights of specific rocks on the Baltic shoreline. His conclusion was often taken in Cuvier's day to be evidence for a slow *worldwide* diminution in sea level (in modern terms, a eustatic change). Only in the mid–nineteenth century was it reinterpreted as a movement of crustal elevation, after further fieldwork showed it was differentially centered in the Gulf of Bothnia; only later still was that movement explained in terms of a slow rebound (in modern terms, an isostatic adjustment) after the weight of the Pleistocene ice sheets over Scandinavia had been removed.]

28. [A reference to suggestions made by both Buffon and Lamarck (the latter in his *Hydrogéologie*, 1802). This and the preceding paragraph were added to the original manuscript text, clearly to forestall possible objections.]

29. [The last two clauses, which were added in manuscript to the original text, probably refer to the celebrated cases of, respectively, the extinct volcanos of central France and a new volcanic island that arose in the Mediterranean in the early eighteenth century. Both were cases in which volcanic action had clearly *changed* in the course of time.]

Thus it would be in vain to seek, among the forces that now act on the earth's surface, causes adequate to produce the revolutions and catastrophes the traces of which are shown in its crust. And if one wants to resort to the constant external causes known at present, one will have no greater success.[30]

[18] ASTRONOMICAL CAUSES

The earth's pole moves in a circle around the pole of the ecliptic; and its axis is inclined more or less to the plane of the ecliptic. But these two movements, the causes of which are now understood, do not surpass certain limits, which are too confined for the effects we have noted. Besides, these excessively slow movements cannot explain catastrophes, which must necessarily have been sudden.

The same reasoning applies to all the slow actions that have been imagined—doubtless in the hope that it would be impossible to deny their existence, because it could always be claimed that their very slowness renders them imperceptible. It matters little whether they are true or not; they explain nothing, since no slow cause can have produced sudden effects. Had there thus been a gradual diminution of the waters, had the sea transported and delivered solid matter, had the temperature of the globe diminished or increased—all that is nothing to what has overturned our strata, enveloped in ice large quadrupeds complete with their flesh and skin, brought onto dry land shellfish as well preserved as if they had been fished out alive, and, finally, destroyed whole species and genera.

These themes have struck the greatest number of naturalists; and among those who have sought to explain the present state of the globe there have been hardly any who have attributed them wholly to slow causes, still less to causes operating under our eyes. The necessity they have felt, to look for causes different from those we see acting today, is also that which has led them to imagine so many extraordinary suppositions, and to err and lose themselves in so many contrary ways, that—as I have said elsewhere—even the name of their science [i.e. geology] has become almost ridiculous for some prejudiced persons, who only see the systems it has hatched, and who forget the long and important series of established facts that it has made known.[31]

30. [The word "constant" was added in manuscript to the original text: a crucial modification, because it served to exclude consideration of the exceptional cosmic events, notably major cometary impacts, that had figured prominently in many of the "theories of the earth" that Cuvier so despised. Such events were of course suspect as explanatory resources, precisely because they were *not* known present-day "causes."]

31. When formerly I mentioned this circumstance, of the science of geology having become ridiculous [see text 13], I only expressed a well-known truth, without presuming to give my own

[19] FORMER SYSTEMS OF GEOLOGISTS

For a long time, only two events — only two epochs of change in the globe — were admitted: the creation and the deluge. All the efforts of geologists strove to explain the present state by imagining a certain primitive state later modified by the deluge, the causes, action, and effects of which were also imagined by each in his own way.

Thus according to one,[32] the earth had at first received a smooth and even crust, which covered the abyss of the oceans, and which on cracking produced the deluge; its debris formed the mountains. According to another,[33] the deluge was occasioned by a momentary suspension of the cohesion of minerals; the whole mass of the globe was dissolved, and the [resultant] paste was penetrated by shells. According to a third,[34] God raised the mountains in order to let the waters of the deluge flow away, and took them into the areas where there were the most rocks, because otherwise they would not have been able to stand up to the flood. A fourth[35] created the earth with the atmosphere of one comet, and flooded it with the tail of another; the heat that remained from its origin was, according to him, what excited all living beings to sin; and they were all drowned, except for the fish, which apparently had less lively passions.[36]

It can be seen that, while taking refuge within the limits set by Genesis, naturalists still retained a fairly vast arena [carrière]. Soon they found themselves cramped; and when they had succeeded in envisaging the six days of creation as so many indefinite periods, centuries no longer cost them anything, and their systems soared in proportion to the spaces that they had at their disposal.

The great Leibniz himself, like Descartes, amused himself by making the

---

opinion, as some respectable geologists seem to have believed. If their mistake arose from my expressions having been rather equivocal, I take this opportunity of explaining my meaning. [Cuvier had modified the passage in manuscript: by adding the qualification that follows the word "ridiculous," he distanced himself from the criticism and conceded that there was *some* evidential value in these speculative "theories of the earth."]

32. Burnet, *Sacred theory of the earth*, London, 1681 [1st ed. 1680]. [The English scholar Thomas Burnet (c. 1635–1715) had been chaplain to King William III in London.]

33. Woodward, *Essay towards the natural history of the earth*, London, 1702 [1st ed. 1695]. [The London physician and naturalist John Woodward (1665–1728) had bequeathed his fine collection of fossils to Cambridge, and had endowed a professorship to promote his theories after his death.]

34. Scheuchzer, *Memoirs of the Academy [of Sciences]*, [for] 1708. [Johann Scheuchzer (1672–1733) had been a prominent Swiss naturalist.]

35. Whiston, *A new theory of the earth*, London, 1708 [1st ed. 1696]. [William Whiston (1667–1752) had been Isaac Newton's successor at Cambridge.]

36. [A nice example of Cuvier's sardonic and irreverent sense of humor, which marks him, if somewhat out of time, as a man of the Enlightenment. The ironic tone of this whole review of earlier "theories of the earth" is unmistakable.]

earth an extinct sun,[37] a vitrified globe, on which vapors, having condensed
at the time that it cooled, formed the seas, and then deposited the limestone formations. De Maillet covered the entire globe with water for thousands of years; he had the waters retiring gradually; all terrestrial animals
were at first marine; man himself had begun as a fish. The author claimed
that it was not uncommon to find in the ocean fish that had as yet only become half human, but of which the race would one day become fully human.[38] Buffon's system is little more than a development of Leibniz's, with
the sole addition of a comet, which by a violent collision caused the liquid
mass of the earth to leave the sun, at the same time as that of all the planets. From this [point], positive dates are derived: for, from the present temperature of the earth one can know how long ago it cooled; and since the
other planets left the sun at the same time, one can calculate how many
more centuries the large ones will take to cool, and to what point the small
ones are already frozen.[39]

[20] MORE MODERN SYSTEMS [40]

In our own day too, minds *[esprits]* more liberal than ever have wanted to
exercise themselves on this great subject. Some writers have reproduced
and prodigiously extended de Maillet's ideas.[41] They say that everything
was fluid in origin; that the fluid generated animals that at first were very
simple, such as monads or other microscopic species of infusoria; that in
the course of time and by taking up diverse habits *[habitudes]*, the races of
these animals became more complex, and diversified to the point where we
see them today.[42] These are all the races of animals that have gradually

37. Leibniz, *Protogaea* [in] *Acta Lipsiae*, 1693; *[Protogaea,]* Göttingen, 1749. [Only a brief summary of Leibniz's theory was published in his lifetime; the full version appeared long afterward.]
   38. *Telliamed* [1748; the title of the anonymous work was an inversion of the name of its author,
de Maillet].
   39. [Buffon, "Époques de la Nature" (1778). Cuvier's summary of Buffon's attempts to give precise dates to these remote events is of course ironic.]
   40. [This section heading, which is required by the sense of the text, is found in the manuscript
but was omitted—probably by mistake—from the published version.]
   41. [In the manuscript this sentence reads, "Some writers in Germany have recently reproduced
the speculations of de Maillet." This suggests that Cuvier originally intended to make the allusion to
Lamarck's work even more oblique than it appeared in the published text (see below).]
   42. [In the manuscript the sentence continues in an important passage that was evidently deleted
in press, probably because Cuvier decided—or was persuaded—that such sarcasm would be im-
prudent or counterproductive: "that the habit of chewing, for example, resulted at the end of a few
centuries in giving them teeth; that the habit of walking gave them legs; ducks by dint of diving became pikes; pikes by dint of happening upon dry land changed into ducks; hens searching for their
food at the water's edge, and striving not to get their thighs wet, succeeded so well in elongating their
legs that they became herons or storks. Thus took form by degrees those hundred thousand diverse
races, the classification of which so cruelly embarrasses the unfortunate race that habit has changed

converted the seawater into calcareous earth; plants—about the origin and metamorphoses of which we are told nothing—have for their part converted the water into clay; but these two earths, as a result of being deprived of the characters that life had imprinted on them, are resolved in the last analysis into silica; and that is why the oldest formations *[montagnes]* are more siliceous than the others. Thus all the solid parts of the earth owe their birth to life; without life, the globe would still be entirely liquid.[43]

Other writers have given their preference to Kepler's ideas. Like that great astronomer, they assign vital faculties to the globe itself; a fluid circulates in it, according to them, and digestion takes place there just as in organisms. Each of its parts is alive; there is not the most elementary molecule that does not have an instinct, a will, and that does not attract or repel according to sympathies and antipathies. Each kind of mineral can convert immense formations *[masses]* into its own nature, just as we convert our food into flesh and blood. Mountains are the globe's respiratory organs, and schists its secretory organs: by these it decomposes seawater in order to generate volcanic ejecta. Finally, veins are its decayed parts, the abscesses of the mineral kingdom, and ores *[métaux]* a product of rot and of sickness; and this is why they almost all smell so bad.[44]

It must be admitted that, while we have chosen some extreme examples, not all geologists have taken the boldness of their conceptions as far as those we have just cited; but how much diversity and contradiction still reign, [even] among those who have proceeded with the greatest caution, and who have not looked for means beyond ordinary physics and chemistry![45]

---

into naturalists" (as translated in Burkhardt, *Spirit of system* [1977], p. 199; original text, p. 257 n. 54). This was probably the kind of joke at the transformists' expense—possibly even the very *same* joke—that Cuvier had made during his first lectures on geology (see text 10).]

43. See Rodig, *Physics* [i.e. *Lebende Natur*], Leipzig, 1801, p. 106, and [de Maillet,] *Telliamed*, p. 169. It is Lamarck who has developed this system recently with the most cogency and most sustained sagacity, in his *Hydrogeology* [1802] and *Zoological philosophy* [1809]. [Cuvier originally drafted a further sentence at this point in the text. Although he deleted it before sending the manuscript to press, it is important for what it reveals about his opinion of what Lamarck's work ultimately entailed: "It can be seen that we have here not only geology; there is also an entirely new chemistry, mineralogy, botany, and physiology; and in fact these creators of the earth usually have yet another concern, that of re-creating all the sciences." The comment was not unfair to Lamarck, who was quite explicit about his ambition to reconstitute many if not all the sciences on new foundations.]

44. Mr. Patrin has used much ingenuity *[esprit]* to support this viewpoint, in several articles in the *New dictionary of natural history* [*Nouvelle dictionnaire d'histoire naturelle*, 1802–4]. [Eugène Patrin (1742–1815) was a French mineralogist who had traveled extensively in Russia.]

45. [Cuvier's first draft for the last clause expressed even more clearly the actualistic criterion he had in mind: "who have only used means that are avowed by experience or by everyday observation."]

For one, everything was precipitated successively, and deposited more or less as it still is; but the sea, which covered all, has retreated by degrees.[46] For another, the materials of the mountains are ceaselessly worn down and carried away by rivers; taken to the depths of the sea, they are heated under enormous pressure, and form strata that the heat that hardened them will one day violently elevate again.[47] A third alleges that the liquid divided into a multitude of lakes, placed in an amphitheater one above the other; after depositing our shelly strata they successively broke their dikes and went to fill the ocean basins.[48] For a fourth, on the contrary, tidal waves seven or eight hundred fathoms [about 4,500 feet] high have from time to time taken the seafloor and thrown it up into mountains, or into hills in the valleys, or onto the original plains of the continents.[49] A fifth has the various fragments that compose the earth falling successively from the sky like meteorites, and bearing, as an imprint of their origin, the remains of unknown organisms that they contain.[50] A sixth makes the globe hollow, and puts in it a magnetic core; this shifts from one pole to the other, under the influence *[gré]* of comets, which in turn moves the center of gravity and the mass of water, thus drowning the two hemispheres alternately.[51]

[21] DIVERGENCES OF GEOLOGISTS' SYSTEMS

We could cite another twenty systems just as divergent as these. Make no mistake: it is not at all our intention to criticize these authors. On the contrary, we recognize that these ideas have generally been conceived by spirited and scientific men; they did not ignore the facts at all, and several have even traveled extensively with the intention of examining them.

46. In his *Geology*, de Lamétherie claims crystallization as the principal cause. [The reference is either to de Lamétherie's *Théorie de la terre* (1797) or to his *Leçons de géologie* (1813); the latter, although not yet published, was based on lectures given at the Collège de France as Cuvier's understudy, so Cuvier might have seen it in manuscript or proof form.]

47. Hutton and Playfair, *Illustrations of the Huttonian theory of the earth*, Edinburgh, 1802. [Cuvier's citation of Playfair's book—as if it were by two authors—strongly suggests that he had not seen it, but only heard about it from reviews or oral reports. It was a common opinion at this time that Hutton (in his *Theory of the earth* [1795]), and Playfair after him, had simply put forward yet another speculative "system"; it should be noted that Cuvier mentions no objection to their conception of unlimited geological time, which in any case they shared with Lamarck and many others.]

48. Lamanon, in various parts of the *Journal of physics* [notably in "Fossiles trouvés dans les carrières de Montmartre" (1782)].

49. Dolomieu, ibid. ["Pierres composées" (1791–92)].

50. Messrs. [Karl Wilhelm and Ernst Franz Ludwig] Marschall, in *Researches on the origin and development of the present order of the world [Untersuchungen über den Ursprung und die Ausbildung der gegenwärtigen Anordnung des Weltbäudes]*, Giessen, 1802.

51. Mr. Bertrand, *Periodic renewals of the terrestrial continents [Renouvellements périodiques des continens terrestres]*, Hamburg, 1799.

[22] CAUSES OF THESE DIVERGENCES

So where could such opposition come from, in the solutions of men who start from the same principles in order to resolve the same problem? Could it have been that the conditions of the problem have never been wholly taken into consideration? What has made it remain until this day indeterminate and susceptible of several solutions: all equally good, abstracted from such and such a condition; all equally bad when a new condition comes to be known, or when attention reverts to some known but neglected condition?

[23] NATURE AND CONDITIONS OF THE PROBLEM

To drop this mathematical language, we will say that almost all the authors of these systems considered only certain difficulties that struck them more than others; they were determined to resolve those in a more or less probable manner, but set aside others equally numerous and important. One, for example, saw only the difficulty of getting the level of the seas changed; another, only that of getting all terrestrial substances dissolved in one and the same liquid; yet another, only that of getting animals believed to be tropical to live in the glacial zone. Exhausting their intellectual forces on these questions, they believed that, in imagining any way whatever to answer it, they had done everything. Furthermore, in thus neglecting all the other phenomena, they did not always think even of determining with precision the scale and limits of those they sought to explain.

That is true above all for the Secondary formations *[terrains]*, although they form the most important and most difficult part of the problem. There has almost never been a concern to fix with care the superposition of their strata, nor the relations between these strata and the species of animals and plants whose remains they contain. Are there animals and plants particular to certain beds, which are not found in others? Which are the species that appear first, or those that come later? Do these two kinds of species sometimes accompany each other? Are there alternations in their return, or, in other words, do the first return a second time, and do the second then disappear?[52] Did these animals and plants live in the places where their remains are found, or have they been transported from elsewhere? Do all of them still live somewhere today, or have they been destroyed wholly or in part? Is there a constant relation between the age of the strata and the resemblance or lack of resemblance of the fossils to living organisms? Is there

52. [Significantly, the four preceding questions—in modern terms, highly stratigraphical ones— were added in manuscript to the more biogeographical ones that follow; the former were still perhaps less familiar to Cuvier's habits of mind than the latter.]

a constant relation of climate between the fossils and those living organisms that resemble them most? Can one conclude that the transport of these organisms—if there has been any—was made from north to south, or from east to west, or by scattering [irradiation] and mixing; and can one distinguish the epochs of such transport by the beds that bear traces of them?

What is to be said about the causes of the present state of the globe, if these questions cannot be answered, and if there are not yet sufficient reasons to choose between the affirmative and the negative? For it is only too true that none of these points is yet absolutely beyond doubt; it seems scarcely even to have been considered that it would be good to clarify them before constructing a system.

[24] REASON THAT THE CONDITIONS HAVE BEEN NEGLECTED

The reason for this singularity will be found, if one reflects that geologists were all either cabinet naturalists who had hardly examined the structure of the formations [montagnes] for themselves, or mineralogists who had not studied in sufficient detail the innumerable varieties of animals and the infinite complication of their various parts. The former only built systems; the latter made excellent observations and truly laid the foundations of the science, but they were unable to complete the edifice.

[25] PROGRESS OF MINERAL GEOLOGY

In fact, the purely mineral part of the great problem of the theory of the earth has been studied with admirable care by de Saussure, and since brought to striking development by Mr. Werner and by the numerous and learned students he has trained. For twenty years the first of these celebrated men laboriously traversed the most inaccessible areas, attacking the Alps, as it were, from all sides and by all their passes; and he disclosed for us all the disorder of the Primitive formations [terrains], tracing more clearly the limit that distinguishes them from the Secondary formations. The second, profiting from the numerous excavations made in the country with the oldest mines in the world, has fixed the laws of succession of the formations [couches]; he demonstrated their respective ages and followed each through all its metamorphoses.[53] Positive geology dates from him, and from him alone, in all that concerns the mineral nature of formations. But neither the one nor the other has given to the identification of fossil organic species, in each kind of formation, the rigor that has become necessary since the number of known [fossil] animals has risen so prodigiously.

53. [The allusion was to the ancient mining industry of Saxony. For Cuvier and his contemporaries, the word "metamorphoses" carried none of the meaning of the modern technical term "metamorphism" (first coined by Lyell some twenty years later).]

It is true that other savants have studied the fossil debris of organisms; they have collected them and had them illustrated by the thousand; and their works will be precious collections of material. But they have been more concerned with animals or plants considered as such, than with the theory of the earth; or they regarded these petrifactions or fossils as curiosities rather than as historical documents; or indeed, finally, they contented themselves with partial explanations of the occurrence of each specimen. So they almost always neglected to look for general laws of position or of the relation of the fossils to the formations.

[26] IMPORTANCE OF FOSSILS IN GEOLOGY

Nevertheless, the idea of such research was quite natural. How was it not seen that the birth of the theory of the earth is due to fossils alone; and that without them we would perhaps never have dreamt that there had been successive epochs, and a series of different operations, in the formation of the globe? In effect, they alone provide the certainty that the globe has not always had the same crust [enveloppe], because it is certain that they would have had to live at the surface before they were buried at depth. It is only by analogy that the conclusion that fossils furnished directly in the case of the Secondary formations [terrains] has been extended to the Primitive ones; and if there had only been formations without fossils, no one would have been able to maintain that those formations had not been formed all together [i.e. at the same period].

Again, it is through fossils—slight though knowledge of them has remained—that we have recognized the little that we do know about the nature of the revolutions of the globe. They have taught us that the formations [couches], or at least those that contain them, were deposited tranquilly in a liquid; that their variations corresponded to those of the liquid; that their exposure [mise à nu] was occasioned by the removal of that liquid; and that that exposure took place more than once. None of all that would have been certain without fossils.

However, the study of the mineral part of geology, although no less necessary, and indeed of even greater utility for the practical arts, is much less instructive in connection with the object here in question. We are absolutely ignorant about the causes of the variations in the substances of which the formations are composed; we do not even know the agents that have been able to hold some of them in solution; and for many of them it is still being disputed whether they owe their origin to water or heat [feu]. Basically it could have been seen previously that there is agreement only on one point, namely, that the sea has changed its position. And how is that known, if not by fossils? Fossils, which gave birth to the theory of the earth, have thus furnished it at the same time with its principal enlightenment *[lumières]*, the only evidence that has hitherto been generally recognized.

It is this idea that encouraged us to be engaged in it; but this field is immense, and a single person could scarcely graze a small part of it. It was therefore necessary to make a choice, and we soon made it. The class of fossils that is the object of this work attracted us from the first, because we saw that it is at once more fertile in precise consequences, yet less well known and richer in new subjects for research.[54]

[27] SPECIAL IMPORTANCE OF THE FOSSIL BONES OF QUADRUPEDS

It is in fact clear that, for several reasons, the bones of quadrupeds can lead to more rigorous results than the remains of any other organisms.

First, they characterize in a clearer manner the revolutions that have affected them. Shells do indeed indicate that the sea existed where they were formed; but their changes of species could at a pinch have come from slight changes in the nature or just the temperature of the liquid. They could also have been related to other accidental causes. Nothing can tell us that, in the depths of the sea, certain species or even genera, after occupying specific areas for a more or less long time, may not have been driven out by others. Here, by contrast, all is precise: the appearance of the bones of quadrupeds, and above all of whole corpses in the beds, indicates either that the bed that bears them was itself at one time exposed *[à sec]*, or at least that there was dry land in its vicinity. Their disappearance renders it certain that this bed had been submerged, or that the dry land had ceased to exist. Thus it is by them that we learn, in a confident manner, the important fact of the repeated irruptions of the sea, which the marine fossils and other products by themselves would not have taught us. And it is by their deeper study that we can hope to recognize the number and epochs of these irruptions.

Second, the nature of the revolutions that have altered the surface of the globe must have had a more thorough effect on terrestrial quadrupeds than on marine animals. Since these revolutions largely consisted of displacements of the seabed, and since the waters must have destroyed all the quadrupeds that they reached, if their irruption was general, they could have made the whole class [of the quadrupeds] perish; or, if they extended only to certain continents at a time, they could at least have annihilated the

---

54. This work shows in fact to what extent this matter was still new, notwithstanding the excellent research of the Campers, Pallases, Mercks, Sömmerrings, Rosenmüllers, Fischers, Faujases, and other savants whose works I have taken great care to cite in those of my chapters to which they are related. [Cuvier listed each name in plural form, as if to suggest a host of other naturalists for whom these could stand as examples (though there were in fact two relevant Campers, father and son).]

species peculiar to those continents, without having the same influence on marine animals. On the contrary, millions of aquatic individuals could have been left high and dry, or buried under new strata, or thrown violently against the shore—and nonetheless their race could be preserved in some more tranquil places, from which it would be propagated anew after the agitation of the seas had ceased.

Third, this more thorough effect is also easier to grasp, and it is easier to demonstrate its effects. Since the number of quadrupeds is limited, and most of their species—at least the large ones—are known, there are greater means to check whether fossil bones belong to one of them, or whether they come from a lost species. By contrast, we are far from knowing all the shellfish and all the fish in the sea, and we are probably still ignorant of the greatest part of those that live in the depths. It is thus impossible to know with certainty whether a species that is found fossilized exists somewhere alive. We also see savants persisting in giving the name "pelagic shells" (that is, shells of the open sea) to belemnites, ammonites, and other genera that have so far only been seen in the ancient formations; intending by that to claim that, if those fossils have not yet been discovered in the living state, it is because they live at depths inaccessible to our nets.

Naturalists have doubtless not yet explored all the continents, and do not even know all the quadrupeds that live in the countries they have traversed. New species of this class are discovered from time to time; and those who have not examined attentively all the circumstances of these discoveries might be able to believe that the unknown quadrupeds whose bones are found in our strata may remain hitherto hidden on islands that have not been encountered by sailors, or in some of the vast deserts that occupy the middle of Asia, Africa, both Americas, and New Holland [Australia].

[28] THERE IS LITTLE HOPE OF DISCOVERING
NEW SPECIES OF LARGE QUADRUPEDS

However, if one examines closely what kinds of quadruped have been discovered recently, and in what circumstances they were discovered, it will be seen that there remains little hope of finding one day those that we have so far seen only as fossils.

Islands of moderate size, situated far from large landmasses, have very few quadrupeds, and most of them are very small; when they have large ones, it is because they have been carried there from elsewhere. Bougainville and Cook found only pigs and dogs on the South Sea islands; the largest quadrupeds on the Antilles were agoutis.

It is true that the large landmasses, such as Asia, Africa, the two Americas, and New Holland, have large quadrupeds, and generally of species peculiar to each; inasmuch that every time these lands were found to have been kept isolated from the rest of the world by their situation, the class of quadrupeds there was found to be entirely different from that which existed elsewhere. Thus when the Spaniards traveled for the first time to South America, they found there not a single one of the quadrupeds of Europe, Asia, or Africa. The puma, jaguar, tapir, capybara, llama, vicuña, and all the capuchin monkeys were for them entirely new creatures, of which they had no idea. The same phenomenon was repeated in our own day when the coasts of New Holland and the adjacent islands began to be examined. The various kangaroos, wombats, dasyures, bandicoots, flying phalangers, platypuses, and echidnas have come to astound naturalists by the strange conformations that break all the rules and slip through all systems [of classification].

So if some large continent remained to be discovered, one could still hope to find some new species, among which could be some more or less similar to those whose remains are disclosed in the bowels of the earth. But a glance at the world map, to see the innumerable directions in which navigators have traversed the oceans, is enough to conclude that there can be no further large landmass, unless it is toward the south pole, where the ice would not allow any form of life to subsist.[55]

Thus it is only from the interior of the earth's large landmasses that unknown quadrupeds can still be awaited. But on a little reflection it will soon be seen that the expectation is scarcely better founded on that side than on that of islands. Doubtless the European traveler cannot easily traverse the vast tracts of country, deserted or supporting only savage tribes; and that is above all true of Africa. But nothing prevents animals from moving in every direction around these countries and going toward the coasts. When there are large mountain chains between the coasts and the deserts of the interior, they will always be interrupted at some points to let the rivers flow through; and in these burning deserts the quadrupeds prefer to follow the banks of the rivers. The coastal tribes also move up the rivers, and promptly come to know—either on their own, or by commerce with and from the traditions of tribes further up—all the noteworthy organisms [productions] that live right up to the sources.

At no epoch, therefore, was a very long time necessary before the civilized nations that frequented the coasts of a large landmass would know its larger animals, or those that were striking in configuration, fairly well. The

55. [Although the Antarctic ice had been sighted many times by explorers, it was still uncertain whether there was any substantial landmass there, or merely (as in the Arctic) a large area of floating ice. Cook's discovery that New Zealand was not a peninsula but a group of islands had finally eliminated any chance of there being a large *temperate* southern continent.]

known facts match this reasoning. Although the ancients never crossed the Imaus[56] and the Ganges in Asia, and had not been very far into Africa beyond the Atlas, they genuinely knew all the large animals of those two parts of the world; and if they did not distinguish all the species, it was their similarity that confused them, not that they had been unable to see them or hear them spoken about.

They knew the elephant, and the [natural] history of that quadruped is more exact in Aristotle than in Buffon. They were not even ignorant of some of the differences that distinguish the African elephants from the Asian.[57] They knew the two-horned rhinoceros: Domitian had one shown [in the circus] in Rome and engraved on his coins, and Pausanias described it very well. The one-horned rhinoceros, faraway though its native country is, was equally well known: Pompey had one shown in Rome, and Strabo accurately described another in Alexandria.[58] The hippopotamus was not as well described as the preceding species; but very exact illustrations of it are found on monuments made by the Romans to represent matters concerning Egypt, such as the Nile statue, the mosaic at Palestrina, and a large number of coins. In fact the Romans saw them several times: Scaurus, Augustus, Antoninus, Commodus, Heliogabalus, Philip,[59] and Carinus[60] [all] showed them.

The two species of camel, the Bactrian and the Arabian, were already very well described and characterized by Aristotle.[61] The ancients knew the giraffe, or camel-leopard; one was even seen alive in the circus in Rome, during the dictatorship of Julius Caesar, in the Roman year 708; and ten were assembled by Gordian III, and killed during Philip's centennial games.[62] If one reads closely the descriptions of the hippopotamus given by Herodotus and Aristotle, which are thought to have been borrowed from Hecataeus of Miletus, one will find that they must have been put together from the descriptions of two different animals, one of which was perhaps the true hippopotamus, and the other certainly the gnu (*Antilope gnu* Gmelin). The Ethiopian boar of Agatharchides, which had horns, was in fact our present-day Ethiopian boar, whose enormous tusks deserve the name of horns almost as much as those of the elephant.[63] The bubal and

56. [A mountain range mentioned by writers in antiquity, variously identified in Cuvier's day as the Caucasus, Urals, or Himalaya.]

57. See my "[Natural] history of elephants" in the second volume [of *Ossemens fossiles* (see the excerpt in text 11, and the first version in text 3)].

58. See my "[Natural] history of the rhinoceros" in the second volume [of *Ossemens fossiles.*]

59. See my "[Natural] history of the hippopotamus" in the second volume [of *Ossemens fossiles.*]

60. Calpurnius, *Eclogae* VI, 66. [The names are those of Roman emperors.]

61. *[Natural] history of animals*, book 2, chap. 1.

62. Julius Capitolinus, *Three Gordians*, chap. 23.

63. Aelian, *On the nature of animals*, V, 27.

the water buffalo were described by Pliny, the gazelle by Aelian, the oryx by Oppian, and the axis [deer] since the time of Ctesias.

Aelian described yaks very well, under the name of the ox with a tail that served as a flyswatter.[64] The buffalo had not been domesticated by the ancients; but the Indian ox spoken of by Aelian,[65] which had horns large enough to hold three amphorae, was indeed the variety of [water] buffalo called arni. The ancients knew of hornless oxen;[66] the African oxen whose horns, attached solely to the skin, are shed with it;[67] the Indian oxen, as swift in a race as horses;[68] those no larger than a goat;[69] the sheep with a broad tail;[70] and those of the Indies, as large as donkeys.[71] The indications given by the ancients about the aurochs, the reindeer, and the elk are all mixed up with fables, but they also prove that they had some knowledge of them; however, since this knowledge was based on reports of uncivilized [grossiers] peoples, it has not yet been submitted to judicious criticism.

Even the polar bear was seen in Egypt under the Ptolemys.[72] Lions and panthers were common at the games in Rome, and were seen by the hundred; even some tigers were seen, and the striped hyena and the Nile crocodile appeared. In the ancient mosaics preserved in Rome there are excellent portraits of the rarest of these species: among others, the striped hyena is represented perfectly in a piece preserved in the Vatican museum; and while I was in Rome a mosaic pavement of natural stones (done in the Florentine manner) was discovered in a garden beside the Arch of Galienus, representing four Bengal tigers finely portrayed.

The Vatican museum possesses a crocodile in basalt, of almost perfect precision.[73] It can hardly be doubted that the "hippotiger" was the zebra, which, however, comes only from the southern parts of Africa.[74] It would be easy to show that almost all the species of monkeys that are at all remarkable, were fairly distinctly pointed out by the ancients, under the

64. Idem, XV, 14.
65. Idem, III, 34.
66. Idem, III, 53.
67. Idem, II, 20.
68. Idem, XV, 24.
69. Idem, ibid.
70. Idem, III, 3.
71. Idem, IV, 32.
72. Athenaeus, [Deipnosophistae], book 5.
73. The only error is one claw too many on the hind foot. Augustus had shown thirty-six [crocodiles]. Dion [Cassius, Historiae Romanae], book 55.
74. Caracalla killed one of them in the circus. Dion [Cassius, Historiae Romanae], book 77. See Gisbertus Cuperus, On elephants depicted on coins [De elephantis in nummis obviis, 1719], part 2, chap. 7.

names of pitheci, sphinxes, satyri, cephi, cynocephali, and circopitheci.[75]

They knew and described rodents, down to quite small species, when they had some conformation or property that was notable.[76] But the small species have no bearing on our subject. It is sufficient to have shown that all the large species that are remarkable in any way, which we know today in Europe, Asia, or Africa, were already known to the ancients. From this we may readily conclude that if they made no mention of small ones, or if they failed to distinguish those that resemble each other too much, such as the various gazelles and others, they were held back by a lack of attention or of method, rather than by obstacles of climate. We conclude likewise that if eighteen or twenty centuries, and the circumnavigation of Africa and the Indies, has added nothing of this kind to what the ancients have taught us, there is no likelihood that the coming centuries will teach much to our descendants [neveux].

Perhaps, however, someone will mount the converse argument, and say that the ancients not only knew as many large animals as we do—as we have just shown—but that they described several that we have not; that we are too hasty in regarding these animals as fabulous; that we ought to go on looking for them before we believe we have exhausted the [natural] history of the existing creation; and finally that among these putatively fabulous animals—when we know them better—will perhaps be found the originals of our bones of unknown species. Some will even think that these various monsters, which are essential ornaments of the heroic history of almost all peoples, are precisely those species that it was necessary to destroy, in order to allow civilization to be established. Thus the Theseuses and Bellerophons would have been happier than all our peoples today, who have indeed driven back the harmful animals, but who have not yet succeeded in exterminating any of them.[77]

It is easy to reply to this objection by examining the descriptions of these unknown beings, and by tracing them to their origin. The most numerous have a purely mythological origin, of which their descriptions bear the unquestionable stamp. For in almost all of them one can see only the parts of known animals, united by an imagination without constraints and against all the laws of nature.

Those that the Greeks invented or put together have at least some grace

75. See Lichtenstein, *Commentary on the forms of monkey that were noticed by the ancients [Commentatio de simiarum quotquot veteribus innotuerunt]*, Hamburg, 1791.

76. The jerboa is engraved on coins of Cyrene, and noted under the name of bipedal rat.

77. [A sarcastic joke at the expense of those whom Cuvier was about to criticize; the mythical Greeks successfully slew the *last* of their respective monsters (their names are given in plural form, to make them representatives of all such hero figures).]

in their composition. Like the arabesques that decorate some of the remains of ancient buildings, and which Raphael's fertile brush has multiplied, the forms that are combined offer the eye some attractive contours,
although repugnant to reason. They are the products of lighthearted dreams,
or perhaps emblems in the oriental style, in which metaphysical or moral
propositions were put forward, veiled under mystical images. Let us forgive
those who spend their time discovering the wisdom hidden in the sphinx of
Thebes, the pegasus of Thessaly, the minotaur of Crete, or the chimera of
the Epirus; but let us hope that no one will seriously search for them in nature: it could be as worthwhile to search there for the animals of Daniel or
the beast of the Apocalypse.

Let us not search further there for the mythological animals of the Persians, which are the progeny of a still more exalted imagination: the *manti-
chore* or "destroyer of men," which carries a human head on a lion's body,
terminating in a scorpion's tail;[78] the *griffin* or "guardian of treasures," half
eagle, half lion;[79] or the *cartazonon*[80] or wild ass with a forehead armed with
a long horn [i.e. unicorn]. Ctesias, who has taken these animals as existing,
has been taken by many authors to be an inventor of fables, although he
had only attributed reality to some hieroglyphic figures. These fantastic
compositions were recovered among the ruins of Persepolis,[81] but what do
they signify? We will probably never know; but they definitely do not represent real beings.

Agatharchides, the other animal fabricator, probably drew on an analogous source. The monuments of Egypt show us numerous further combinations of the parts of different species: men with animal heads and animals
with human heads, which have produced the cynocephali, sphinxes, and
satyrs. The custom of representing men of very different size in one and the
same scene—the king or victor gigantic, subjects or the vanquished three or
four times smaller—would have given rise to the fable of the pygmies. It is
in some nook of one of these monuments that Agatharchides would have
seen his carnivorous bull, whose mouth, split right to the ears, would spare
no other animal;[82] but assuredly no naturalist will accept it, for nature does
not combine cutting teeth with either cloven feet or horns.

78. Pliny, *[Naturalis historiae]*, VIII, 21; Aristotle, *[Historia animalium]*; Photius, *Library* [of earlier works], art. 72; Ctesias, *On India*; Aelian, *[Nature of] animals*, IV, 21.

79. Aelian, *Animals*.

80. Aelian, *Animals*, XVI, 20; Photius, *Library*, art. 72; Ctesias, *India*.

81. See Corneille le Brun [Lebrun], *Travels in Muscovy, Persia, and the East Indies* [1718], vol. 2;
and the German work of Mr. Heeren, on the commerce of the ancients [*Geschichte der Staaten des
Alterthums*, 1810].

82. Photius, *Library*, art. 250; Agatharchides, *Historical excerpts*, chap. 39; Aelian, *Animals*, XVII,
45; Pliny, *[Naturalis historiae]*, VIII, 21.

There will perhaps have been many other figures equally strange, either on those [Egyptian] monuments that have not withstood time, or in Arabian and Ethiopian temples that have been destroyed by the religious zeal of Muhammadans and Abyssinians. Those in India swarm with them; but their combinations are too extravagant to have deceived anyone: monsters with a hundred arms or with twenty different heads are also far too monstrous. There is no one, including the Japanese and Chinese, who are without imaginary animals that they claim as real, and that they even illustrate in their religious books. The Mexicans had them: it is the custom among every people, when their idolatry is not yet at all refined. But who would dare claim to find these progeny of ignorance and superstition in nature?

However, what will have happened is that travelers, in order to have themselves valued, will have claimed to have seen these fantastic beings; or, from lack of attention, and deceived by some slight resemblance, they will have taken real beings for these. The great apes will have appeared to be true cynocephali, true sphinxes, true men with tails; thus it is that Saint Augustine will have believed he had seen a satyr.

Some true animals, poorly observed and poorly described, will also have given rise to monstrous ideas, although founded on some reality. Thus the existence of the hyena cannot be doubted, even though this animal does not have a neck supported by a single bone nor change its sex each year, as Pliny said. Thus the carnivorous bull is perhaps just a two-horned rhinoceros with two misshapen horns. Mr. Wertheim claims indeed that Herodotus's gold-bearing ants are corsac foxes.

One of the most famous among these animals of the ancients is the *unicorn*. It has been sought persistently up to our own day, or at least arguments have been sought to support its existence. Three animals are mentioned frequently among the ancients as having only one horn in the middle of the forehead: the *African oryx*, which also has a cloven foot, hair going the wrong way,[83] a large size comparable to that of the ox[84] or even the rhinoceros,[85] and which it is agreed approaches the deer and goats in form;[86] the *Indian ass*, which is single-hoofed; and the *monoceros* proper, whose feet are sometimes compared to those of the lion,[87] sometimes to those of the elephant,[88] and which in consequence is supposed to be cloven-hoofed. The unicorn horse[89] and ox are both doubtless related to the Indian ass, for

83. Aristotle, *Animals*, II, 1, and III, 2; Pliny, XI, 46.
84. Herodotus, *[Historia]*, IV, 192.
85. Oppian, *On the chase, [Cynergetica]*, II, verse 551.
86. Pliny, VIII, 53.
87. Philostorgius, *Ecclesiastical history*, III, 11.
88. Pliny, VIII, 21.
89. Onesicritus, in Strabo, *[Geographia]*, book 15; Aelian, *Animals*, XIII, 42.

even the ox is given as single-hoofed.[90] I ask: if these animals had existed as distinct species, would we not at least have their horns in our museums? And what unpaired horns do we possess, other than those of the rhinoceros and the narwhal?

How then are we to regard the crude figures drawn by savages on rocks? Not knowing about perspective, and wanting to represent a straight-horned antelope in profile, they could give it only a single horn: and there you have an oryx right away. Likewise the oryxes on Egyptian monuments are probably just products of the stiff style imposed on the artists of that country by their religion. Many of their quadrupeds in profile have only one foreleg and one hind; so why would they have shown two horns? Perhaps it happened that individuals were taken that had been deprived of one horn by some accident, as happens quite often in chamois and saigas; that would have sufficed to confirm the error induced by these images. Besides, not all the ancients reduced the oryx to a single horn; Oppian expressly gives it several, and Aelian cites it as having four;[91] and finally, if this animal was a ruminant with cloven hooves, it certainly had a frontal bone divided in two, and—following Camper's well-judged remark—could not have borne a horn on the suture.

But, it will be asked, what two-horned animal could have given the idea of the oryx, by possessing the traits that relate to its conformation—even leaving aside the single horn? With Pallas, I reply that it is the straight-horned antelope, badly named *pasan* by Buffon (*Antilope oryx* Gmelin). It lives in the African deserts and must come as far as the borders of Egypt, and it is what the hieroglyphs seem to represent. Its form is much like that of a stag and its size equal to that of an ox; the hair on its back is directed toward the head; its horns form fearsome weapons, pointed like spears and as hard as iron; its fur is whitish; and its face bears bands and features of black. There you have all that the naturalists have said about it; and as for the fables of the Egyptian priests—which were motivated by the adoption of its image among the hieroglyphic signs—it is not necessary that they were founded on nature. Thus, that an oryx deprived of one horn had been seen, and had been taken as a normal specimen, typical of the whole species; and that that error, adopted by Aristotle, was copied by his successors—all that is possible, even natural, and nonetheless proves nothing about the existence of a single-horned [*unicorne*] species.

As for the Indian ass, when one reads that the ancients regarded its horn as an antidote to poison, it will be seen that this is exactly what orientals

90. Pliny; Solinus, [*Memorabilia mundi*].
91. On [*the nature of*] *animals*, book 15, chap. 14.

today attribute to rhinoceros horn. At the time when this horn would have first been taken to the Greeks, they would not yet have known the animal that bore it. In fact Aristotle does not mention it at all, and Agatharchides was the first to describe it. It is thus that they had ivory long before knowing the elephant. Perhaps some of their travelers would even have called the rhinoceros an "Indian ass," with as much justice as the Romans named the elephant a "Lucanian ox." And all that is said of the strength, size, and ferocity of this wild ass fits the rhinoceros very well. Consequently, those who knew the rhinoceros better, on finding this name "Indian ass" in earlier authors, would have taken it uncritically to be a specific animal; and so from the name it would have been concluded that the animal must be single-hoofed. There is indeed a more detailed description of the Indian ass by Ctesias,[92] but we saw above that this was based on the bas-reliefs at Persepolis, and therefore ought to count for nothing in the positive [natural] history of the animal.

Finally, when we come to some rather more precise descriptions that speak of an animal with a single horn, but with several digits, a third species has been made from them, under the name of "monoceros." This kind of double usage is the more frequent in the ancient naturalists, because almost all those whose works have reached us were simple compilers; because Aristotle himself frequently mixed up facts taken from elsewhere with those he himself had observed; and finally because the art of criticism was then as little known among naturalists as among historians—which is saying a lot.

From all this reasoning and all these digressions, it follows that the large animals that we know in the Old World were [all] known to the ancients; and that the animals described by the ancients and not known in our time were fabulous. It also follows that it did not require a lot of time for the large animals of the three first parts of the world [i.e. Europe, Asia and Africa] to be known to the peoples that frequented their coasts.

It can be concluded from this that there are also no large species to be discovered in America. If they existed there, there would be no reason why we would not know of them; and in fact none have been discovered there in the past 150 years. The tapir, jaguar, puma, capybara, llama, vicuña, red wolf, buffalo or American bison, anteaters, sloths, and armadillos are already [described] in Margrave and Hernandez, as in Buffon;[93] it could even be said that they are better there, for Buffon muddled the [natural] history

92. Aelian, *Animals*, IV, 52; Photius, *[Bibliotheca]*.

93. [Margravius, *Historiae rerum naturalium Brasiliae* (Natural history of Brazil, 1648) and Hernandez, *Historiae animalium et mineralium Novae Hispaniae* (Natural history of animals and minerals of New Spain, 1649) were early scientific accounts of the New World.]

of the anteaters, ignored the jaguar and red wolf, and confused the American bison with the Polish aurochs. In fact Pennant is the first naturalist who has properly distinguished the small musk ox, though it had long been noted by travelers.[94] Molina's horse with cloven hooves was not described at all by the first Spanish travelers; but it is more than doubtful that it exists, and the authority of Molina is too suspect to make one accept it.[95] It can thus be said that the bighorn sheep of the Blue Mountains is up to now the only American quadruped of any size, the discovery of which is fully modern; and perhaps it is just an argali that has come from Siberia over the ice.

How then can it be believed that the huge mastodons and gigantic megatheriums, whose bones are found underground in the two Americas, still live on that continent? How can they have escaped the nomadic peoples that ceaselessly move around the continent in all directions, and who themselves recognize that they [the animals] no longer exist? For they have devised a fable about their destruction, saying that they were killed by the Great Spirit in order to prevent the annihilations of the human race. But one can see that this fable was occasioned by the discovery of the bones, like that of the inhabitants of Siberia, who claim that the mammoth lives underground like a mole, and like all those of the ancients, who identified giants' tombs wherever they found elephant bones.

Thus it can well be believed that if—as we shall say shortly—none of the large species of quadrupeds now buried in the regular rocky strata has been found to be similar to known living species, it is not the effect of simple chance, nor because just those species of which we have only the fossil bones are hidden [alive] in deserts and have hitherto escaped the notice of all travelers; but that the phenomenon must be regarded as being due to general causes,[96] and its study as one of the most suitable to enable us to infer the nature of those causes.

[29] THE FOSSIL BONES OF QUADRUPEDS ARE DIFFICULT TO IDENTIFY

But if this study is more satisfactory in its results than the study of other remains of fossil animals, it also bristles with much more numerous difficulties. Fossil shells are ordinarily preserved complete, with all the characters that allow them to be recognized in collections or in the works of naturalists. Even [fossil] fish offer more or less complete skeletons; the general form of the body can almost always be distinguished, and most often their

94. [Thomas Pennant (1726–98) had been a prominent English naturalist and traveler.]

95. [Juan Ignacio Molina (1740–1829).]

96. [That is, natural processes, not human agency. Cuvier first wrote, "Due to large-scale causes that have brought the surface [enveloppe] of the globe into its present state."]

generic and specific characters, which are normally based on their hard parts. Among quadrupeds, on the contrary, when a complete skeleton is met with, it is difficult to apply to it characters drawn for the most part from the skin, from colors, and from other marks that vanish before fossilization. It is also extremely rare to find a fairly complete fossil skeleton. Isolated bones, scattered higgledy-piggledy, and almost always broken and reduced to fragments, are all that our beds give us in this class, and are the naturalist's only resource. Furthermore, it can be said that the majority of observers, scared off by these difficulties, have skated lightly over the fossil bones of quadrupeds: they have classified them in a vague way, by superficial resemblances, or they have not even ventured to give them a name. As a result, this part of the [natural] history of fossils, the most important and instructive of all, is also the least cultivated of all.[97]

[30] PRINCIPLES OF IDENTIFICATION

Fortunately, comparative anatomy possessed a principle that, when well developed, was capable of making these obstacles vanish. It was that of the correlation of forms in organized beings, by means of which each kind of being could be recognized, at a pinch, from any fragment of any of its parts.

Every organized being forms a whole, a unique and closed system, in which all the parts correspond mutually, and contribute to the same definitive action by a reciprocal reaction. None of its parts can change without the others changing too; and consequently each of them, taken separately, indicates and gives all the others.

Thus, as I have said elsewhere, if the intestines of an animal are organized in such a way as to digest only flesh—and fresh flesh—it is also necessary that the jaws be constructed for devouring prey; the claws, for seizing and tearing it; the teeth, for cutting and dividing its flesh; the entire system of its locomotive organs, for pursuing and catching it; its sense organs, for detecting it from afar; and it is even necessary that nature should have placed in its brain the instinct necessary for knowing how to hide itself and set traps for its victims. Such are the general conditions of the carnivorous regime; every animal adapted [disposé] for this regime unfailingly combines them, for its species could not have subsisted without them. But within these general conditions there exist particular conditions, relative to the size, species, and habitat [séjour] of the prey to which the animal is adapted;

97. As I have already written above [note 54], I do not claim by this remark to detract from the value of the observations by Messrs. Camper, Pallas, Blumenbach, Sömmerring, Merck, Faujas, Rosenmüller, etc.; but their admirable works—which have been very useful to me, and which I cite throughout—are only partial [partiel: i.e. either incomplete, or local, or both].

and each of these particular conditions results in some detailed circumstances in the forms that result from the general conditions. Thus it is not only the class that finds expression in the form of each part, but the order, the genus, and even the species.

In fact, for the jaw to be able to seize [the prey], it must have a certain form of condyle; a certain relation between the position of the resistance and the [muscular] power with the point of support; a certain volume in the temporal muscles, which demand a certain size to the fossa that accommodates them, and a certain convexity to the zygomatic arch under which they pass; and that zygomatic arch must also have a certain strength to give support to the masseter muscle. For the animal to be able to carry its prey, it must have a certain strength in the muscles that raise its head, from which results a specific form of the vertebrae where the muscles have their attachment and in the occiput where they are inserted. For the teeth to be able to cut flesh, it is necessary that they be sharp; and that they be more or less so, according as they are used to cut flesh more or less exclusively. Their base has to be the more solid, the more they have to cut bones, and the thicker those bones are. All these circumstances also influence the development of all the parts that serve to move the jaw.

For the claws to be able to seize the prey, a certain mobility of the digits will be necessary,[98] and a certain strength in the claws; this will result in specific forms in all the phalanges, and in the necessary distribution of the muscles and tendons. It will be necessary for the forearm to have a certain facility for turning, which results again in specific forms for the bones that it comprises. But the bones of the forearm, articulating with the humerus, cannot change in form without entailing changes in the latter. The shoulder bones will have to have a certain degree of firmness in animals that use their forelimbs for grasping, and that will result again in particular forms for them. The play of all these parts will require certain proportions in all their muscles; and the impressions [i.e. attachments] of these muscles being thus proportioned, will determine even more particularly the forms of the bones.

It is easy to see that similar conclusions can be drawn for the hindlimbs, which contribute to the rapidity of movement in general; for the composition of the trunk and the forms of the vertebrae, which influence the facility and flexibility of those movements; and for the form of the nasal bones, the eye socket, and the ears, for which the connections with the perfection

---

98. [The striking change to the future tense suggests an intensification of Cuvier's active role in the imaginative construction of the demands of the carnivorous mode of life.]

of the senses of smell, sight, and hearing are obvious. In a word, the form of the tooth entails *[entraîne]* the form of the condyle; the forms of the shoulder blade and the claws, just like the equation of a curve, entail all their properties. Just as in taking each property separately as the basis for a particular equation, one would find both the ordinary equation and all the other properties of any kind, so likewise the claw, the shoulder blade, the condyle, the femur, and all the other bones taken separately, determine *[donnent]* the teeth, and each other reciprocally.[99] Beginning with each of them in isolation, he who possesses rationally the laws of organic economy would be able to reconstruct the whole animal.

In a general sense, this principle is sufficiently clear in itself not to need any fuller demonstration. But when it comes to applying it, there is a large number of cases in which our theoretical knowledge of the relationships of the forms would be insufficient, if they were not founded on observation.

For example, we see clearly that hoofed animals must all be herbivores, since they have no means of seizing prey. We also see that, as their forefeet have no other use than to support the body, they do not need such a robustly organized shoulder [as a carnivore], from which results the absence of the clavicle and acromion, and the narrowness of the shoulder blade. Not having any need to twist their forearms either, their radius will be fused to the cubitus, or at least joined with gynglymus to the humerus. Their herbivorous regime will require teeth with a flat crown, to grind seeds and grasses. That crown must be uneven; and to be so, the enamel parts must alternate with bony parts. Since this kind of crown necessitates horizontal movements for grinding, the condyle of the jaw cannot be a hinge as tight as in the carnivores: it will have to be flattened, and to match a more or less flattened facet in the temporal bone. The temporal fossa, which will have to accommodate only a small muscle, will have little width or depth, etc. All these things can be deduced one from the other, according to their greater or lesser generality, given that some are essential and belong exclusively to hoofed animals, while others, although equally necessary for these animals, will not be exclusive to them but could also be found in other animals in which other conditions still allow them.

If one then descends to the orders or subdivisions of the class of hoofed animals, and one examines which modifications the general conditions undergo—or rather, which particular conditions are added in, according to the distinctive character of each order—the reasons for these subordinate

---

99. [A bold analogy with mathematics, implying a claim to the kind of prestige enjoyed by the sciences practiced by Cuvier's patron Laplace.]

conditions begin to seem less clear. One can indeed also broadly conceive that a more complicated digestive system is necessary in species in which the dental system is less perfect. Thus it can be said that such species ought rather to be ruminants, in which such and such an order of teeth is lacking; a certain form of esophagus and corresponding forms for the neck vertebrae could be deduced, and so on. But I doubt if it would have been guessed—if it had not been learned by observation—that all ruminants would have a cloven hoof, and they alone; or that only in this class would there be horns on the forehead; or that those of them that have pointed canines would be the only ones to lack horns, and so on.

Nevertheless, since these relations are constant, they must have a sufficient cause. But as we do not know it, where theory fails observation must provide. It establishes empirical laws that become almost as certain as rational ones, when they are based on sufficiently repeated observations. For example, someone today who just sees the track of a cloven hoof can conclude that the animal that left that imprint was a ruminant; and that conclusion is quite as certain as any other in physics or morals. Thus that single track gives the observer the form of the teeth, jaws, and vertebrae, and the form of all the bones of the legs, thighs, shoulders, and pelvis of the animal that just passed by. It is a more certain mark than all those of Zadig.[100]

Observation itself gives a glimpse that there are always hidden reasons for all these relations, independent of general philosophy.

When a general system of these relations is established, one notices not only a specific constancy—if it can be so expressed—between a certain form of a certain organ, and another form of a different organ; but one also perceives a standard constancy, and a corresponding gradation in the development of these two organs, which show their mutual influence almost as well as actual reasoning.

For example, the dental system of nonruminant hoofed animals is generally more perfect than that of cloven-hoofed or ruminant animals, because the former have incisors or canines, and almost always both on both jaws. The structure of the foot is generally more complex, because there are more digits, or a hoof that covers less of the phalanges, or more distinct metacarpal and metatarsal bones, or more numerous tarsal bones, or a fibula more distinct from the tibia; or indeed finally because all these features are often united. It is impossible to give the reasons for these relations. But

100. [In Voltaire's famous story of that name, Zadig successfully traced the king's horse and the queen's bitch, both of which had escaped, by following their tracks. Cuvier uses that fictional achievement as a measure of his own *greater* skill in reconstructing a whole organism and its habits from fragmentary traces.]

what proves they are not at all a product of chance is that every time a cloven-hoofed [animal] shows in the arrangement of its teeth a tendency to approach those we are speaking of, it also shows the same in the arrangement of the feet. Thus the camels, which have canines and even two or four incisors in the upper jaw, have an extra bone in the tarsus, because the scaphoid is not fused to the cuboid; and very small hoofs with corresponding ungual phalanges. The chevrotains [mouse or musk deer], in which the canines are very well developed, have a distinct fibula the whole length of the tibia, whereas the other cloven-hoofed [animals] have as a fibula only a little bone articulated to the base of the tibia. There is thus a constant harmony between two organs that appear completely alien to one another; and the gradation in their forms corresponds without interruption, even in the cases where we cannot give the reason for their relations.

Now in thus adopting the method of observation as a supplementary means, when theory abandons us, we reach details that are astonishing. The least facet on a bone, the smallest apophysis, has a specific character relative to the class, the order, the genus, and the species to which it belongs: to the point that every time one has just a well-preserved extremity of a bone, one can—with application, and the aid of a little appeal to analogy and effective comparison—determine all these things as surely as if one possessed the whole animal. I experimented several times with this method, on portions of known animals, before entirely placing my confidence in it for fossils; but it always succeeded so infallibly that I no longer have any doubt about the certainty of the results it has yielded.

It is true that I have enjoyed all the assistance that could be necessary; and that my fortunate position, and assiduous research spanning nearly fifteen years, have provided me with the skeletons of all the quadruped genera and subgenera, and even, in certain genera, of many of the species and of several individuals in some species. With such resources it has been easy for me to multiply my comparisons, and to verify in all their details the applications that I made of my laws.

We cannot discuss the extent of this method any further, and we must refer to the large comparative anatomy that will soon appear,[101] where all the rules will be found. However, the intelligent reader will already be able to extract a large number of them from the present work, if he takes the trouble to follow all the applications we have made of them here. He will see that it is by this method alone that we have been directed, and that the

101. [That is, his *Règne animal* (Animal kingdom) which was eventually published five years later, in 1817.]

method has almost always been adequate to refer each bone to its species, when it was of a living species; to its genus, when it was of an unknown species; to its order, when it was of a new genus; and finally to its class, when it belonged to an order not yet established. And in the last three cases it has been adequate to assign the characters appropriate for distinguishing it from the orders, genera, or species that most resemble it. Before us, naturalists were not taking advantage of it for whole animals. It is thus that we have identified and classified the [fossil] remains of seventy-eight quadruped animals, both viviparous and oviparous.[102]

[31] TABULATION OF THE RESULTS OF THE PRESENT WORK

Considered in relation to species, forty-nine of these animals were definitely unknown to naturalists until now; eleven or twelve have such an absolute similarity with known species that one can scarcely retain any doubt about their identity; the sixteen or eighteen that remain show many points of resemblance to known species, but it has not yet been possible to make the comparison in a sufficiently scrupulous way to relieve all doubt.

Considered in relation to genera, of the forty-nine unknown species there are twenty-seven that belong to new genera, and these genera are seven in number. The twenty-two other species belong to sixteen known genera or subgenera. The total number of genera or subgenera to which I have assigned fossil bones—of known or unknown species—is thirty-six.

It is not useless to consider fossil animals also in relation to classes and orders. Of the seventy-eight [species], fifteen, in eleven genera or subgenera, are oviparous quadrupeds, and all the others are mammals. Of the latter, thirty-two belong to nonruminant hoofed animals, in ten genera; twelve to ruminants, in two genera; seven to rodents, in six genera; eight to carnivores, in five genera; two to bradypod edentates [three-toed sloths], in only a single genus; and two to amphibians, in two genera.

However, it would still be premature to base on these numbers any conclusion relative to the theory of the earth, because they have no necessary relation at all to the numbers of genera or species that could be buried in our strata [couches]. Thus the bones of large species, which are more striking to workmen, have been collected much more, while those of small species are usually neglected, unless by chance they fall into the hands of a naturalist, or unless some particular circumstance such as their extreme abundance in a certain place draws the attention of even the public [vulgaire].

---

102. [That is, both mammals and reptiles.]

[32] RELATIONS BETWEEN SPECIES AND BEDS [COUCHES]

What is more important—what indeed comprises the definitive object of all my work and establishes its true relation to the theory of the earth—is to know in which beds each species is found, and whether there are some general laws, relative either to the zoological subdivisions or to the greater or lesser resemblance of the species to those of today.

The laws recognized in this respect are very beautiful and very clear. First, it is certain that oviparous quadrupeds appeared much sooner than the viviparous. The crocodiles of Honfleur and England are below the chalk.[103] The monitors of Thuringia are even more ancient, if—as the school of Werner believes—the coppery shales that conceal them (along with so many kinds of what are believed to be freshwater fish) are among the oldest beds of the Secondary formations.[104] (The large saurians [i.e. mosasaurs] and turtles of Maastricht are in the chalk formation itself, but they are marine animals.) This first appearance of fossil bones thus already seems to show that dry land and freshwater existed before the formation of the chalk; but neither at that epoch nor during the formation of the chalk, nor even for a long time after that, were any bones of land mammals fossilized [incrusté].

We begin to find bones of marine mammals, namely those of sea cows and seals, in the coarse shelly limestone that covers the chalk in our [Parisian] environs, but there are still no bones of land mammals. Despite the most persistent research, I have been unable to discover any distinct trace of that class, before the formations deposited on the coarse limestone; but as soon as those formations are reached, the bones of land animals show themselves in great numbers.

Thus, just as it is reasonable to believe that shells and fish did not exist at the epoch of the formation of the primordial rocks [terrains], one should also believe that the oviparous quadrupeds began with the fish, from the first times that produced the Secondary formations; but that the land quadrupeds came only a long time later, when the coarse limestones—which already contain most of our genera of shells, though in species different from ours—had been deposited.

It is worth noting that these coarse limestones, which serve as building stone in Paris, are the last beds to show that the sea stayed long and calmly on our continents. After them one does indeed find further formations

103. [In modern terms these reptiles (not now classed as crocodiles) came from Upper and Lower Jurassic formations respectively.]

104. [These reptiles (not now classed as monitors) came from the *Kupferschiefe*, later assigned a Permian age.]

filled with shells and other marine productions, but these are loose formations of sands, marls, sandstones, and clays, which indicate more or less tu-
multuous transport rather than a quiet precipitation; and if there are some
inconsiderable regular stony beds below or above these detrital formations
[terrains de transport], they generally show marks of having been deposited
in freshwater.

Thus all the known bones of viviparous quadrupeds are either in these
freshwater formations or in these detrital formations; and consequently
there is every reason to believe that these quadrupeds had not begun to exist, or at least to leave their remains in our beds *[couches]*, until after the
penultimate retreat of the sea, and during the state of things that preceded
its last irruption.

But there is also an order within the disposition of these bones themselves, and this order shows a very remarkable succession among their
species.

First, all the genera unknown today—the palaeotheriums, anoplotheriums, etc., about whose [geognostic] position *[gisement]* one can be certain—
belong to the oldest of the formations that are in question here, namely
to those that rest immediately on the coarse limestone. They are principally those that fill the regular beds deposited by freshwater or certain detrital beds; formed very anciently, generally composed of sands and rolled
pebbles, and perhaps the first alluvia of that ancient world. Some lost species of known genera are also found with them—but in small numbers—
and some oviparous quadrupeds and fish, all of which appear to be from
freshwater. The beds that conceal them are always more or less covered by
detrital beds filled with shells and other marine productions.

The most famous of the unknown species that belong to known genera,
or to genera very close to those that are known—such as the elephants, rhinoceros, hippopotamuses, and fossil mastodons—are not found at all with
these older genera. It is only in the detrital formations that they are discovered, sometimes with marine shells, sometimes with freshwater shells,
but never in regular stony beds. All that is found with these species is either
unknown like them, or at least doubtful. Finally, the bones of species that
appear to be the same as ours are unearthed only in the last alluvial deposits formed on the banks of rivers, or on the floors of ancient lagoons or
dried-out marshes, or in the middle of peat beds, or in fissures and caves in
some rocks, or finally close to the surface in places where they could have
been buried by landslides or by the hand of man. And their superficial position is such that these bones, the most recent of all, are also almost always
the least well preserved.

It is not necessary, however, to believe that this classification of the various positions *[gisemens]* is as clear-cut as that of the species, or that it bears a comparable character of demonstration. There are many reasons why it is not so. First, all my identifications of species have been made on the bones themselves, or on good illustrations; by contrast I am far from having observed for myself all the places where these bones have been discovered. Very often I have been obliged to rely on vague and ambiguous accounts, made by persons who themselves did not know what ought to be observed; more often still I have found no information at all.

Second, it is possible to have in this respect infinitely more equivocation than with regard to the bones themselves. The same formation can appear to be recent in areas in which it is superficial, and ancient in those where it is covered by the beds that succeeded it; ancient formations could have been transported by local inundations, and have covered recent bones; they can have collapsed onto them, have enveloped them, and have mixed with the productions of the ancient sea that they previously contained; ancient bones can have been washed out by water and then taken back by recent alluvia; and finally, recent bones can have fallen into fissures or caves in ancient rocks, and have been enveloped there by stalactites or other incrustations. In each case it would be necessary to analyze and evaluate all these circumstances that could mask the true origin of the fossils; and the persons who collected the bones have rarely suspected that necessity, as a result of which the true character of their emplacement is almost always neglected or unrecognized.

Third, there are several doubtful species that will more or less alter the certainty of the results, as long as clear distinctions about them have not been reached. Thus the horses and buffalos that are found with the elephants do not yet have particular specific characters; and geologists who do not wish to adopt my different epochs for fossil bones will still for many years be able to derive from this an argument that is all the more convenient, in that they will be taking it from my book.

But while agreeing that these epochs are susceptible to some objections, for persons who consider casually some particular case, I am no less persuaded that those who embrace the ensemble of the phenomena will not be checked by these little local difficulties, and that they will recognize with me that there has been at least one, and very probably two, successions in the quadruped class before that which today populates the surface of our countries.

Here I still anticipate another objection, and indeed it has already been made to me.

[33] THE LOST SPECIES ARE NOT VARIETIES OF LIVING SPECIES

Why, it will be said, could not the present races be modifications of the former races that are found among the fossils, modifications that would have
been produced by local circumstances and climatic change, taken to this
extreme difference by the long succession of years?

This objection should appear powerful above all to those who believe in
the indefinite possibility of the alteration of the forms of organisms, and
who think that with [the passage of] centuries and [changes of] habits
[habitudes] all species could be changed one into another, or could be derived from a single one of them. However, one could respond—in their own
terms—that if species have changed by degrees one ought to find some
traces of these gradual modifications; that one ought to find some intermediate forms between the palaeotherium and present-day species, and that
up to now that has not happened at all. Why have the entrails of the earth
preserved no monuments of such a curious genealogy, unless it is because
the species of former times were as constant as ours, or at least because
the catastrophe that destroyed them did not leave them time to give effect
to their variations? And to respond to those naturalists who recognize that
variations are restricted within certain limits fixed by nature, it is necessary
to study how far those limits extend: an intriguing kind of research, very
interesting in itself from an infinity of points of view, yet one that has hitherto been given very little attention.

Such research assumes the definition of the species that serves as the
basis for the use that is made of the word, namely that a species comprises
*individuals that are descended one from the other, or from common parents,
and those that resemble such individuals as much as they resemble each
other.* Thus, we call varieties of a species only those more or less different
races that can be derived from a species by reproduction *[génération]*. Our
observations on the differences between ancestors and descendants are thus
for us the only rational rule, for any other would revert to hypotheses without evidence *[preuves]*.

In thus defining a *variety*, however, we observe that its constitutive
differences depend on particular circumstances, and that their extent in-
creases with the intensity of those circumstances. Thus the most superficial
characters are the most variable: color is closely connected to light, thickness of fur to heat, size to the abundance of food. But in a wild animal even
these varieties are strictly limited by its nature; it does not voluntarily
move away from the places where it finds—to an acceptable degree—all
that is needed for the maintenance of its species, and it does not extend far
from where it finds these conditions combined. Thus although the wolf
and the fox live from the tropical to the arctic zone, in all that immense

distance they scarcely sustain any variation other than a little greater or lesser beauty in the fur. I have compared the skulls of foxes from the Arctic [Nord] and from Egypt with those from France, and have found only some individual differences.

Wild animals confined within lesser spaces vary still less, above all among the carnivores. A more bushy mane is the sole difference between the Persian and Moroccan hyenas. Wild herbivorous animals suffer the influence of climate a little more deeply, because it is combined with that of food, which comes to differ both in abundance and in quality. Thus elephants will be larger in one forest than in another; they will have slightly longer tusks in places where their food is more favorable to the formation of the material of ivory (it is the same in reindeer and stags in regard to their antlers); but if one takes the two elephants that are most unlike each other, one sees not the least difference in the number or the articulations of the bones, in the teeth, etc. Moreover, herbivorous species in the wild state seem to be more restricted in their dispersal than carnivores, because the kind of food combines with temperature to limit them.

Nature is also careful to prevent the alteration of species that could result from their mixture, by the mutual aversion she has given them. It requires all kinds of ruse and human constraint to achieve these unions, even between the species that resemble each other the most; and when the offspring are fertile—which is very rare—their fecundity does not extend beyond a few generations, and would probably not take place without the continuation of the care that brought them into being. Likewise, in our woods, we do not see individuals intermediate between the hare and the rabbit, the red deer and the fallow deer, the marten and the weasel.

But the empire of man alters this order; man develops all the variations of which the type of each species is susceptible, and draws out forms that the species, left to themselves, would never have produced.

Here the degree of variation is still proportional to the intensity of their cause, which is that of slavery [esclavage]. It is not very great in semi-domesticated species such as the cat. Softer fur, more vivid colors, and a more or less robust size are all that it experiences; but the skeleton of an Angora cat does not differ in any constant way from that of a wild cat. In domesticated herbivores, which we transport into all kinds of climate, and subject to all kinds of regime (which we measure differentially in work and food), we get greater variations, though they are still all superficial. A greater or lesser size, more or less long horns (sometimes completely lacking), and a more or less pronounced mass of fat on the shoulders: these form the differences among cattle; and these differences are conserved for a long time even in races transported out of the country in which they were developed, if care is taken to prevent their cross-breeding. The innumerable varieties of sheep are also of the same kind; they principally concern the wool, because that is the object to which man has given most attention. They are somewhat fewer in horses, though still very noticeable.

In general the forms of the bones vary little, and their connections and articulations and the forms of the large teeth never vary. The small development of the tusks in the domestic pig, and the fusion of the hoofs in some of its races, are the most extreme differences we have produced in domestic herbivores.

The most marked effects of human influence are shown in the animal that man has conquered most completely, namely the dog—that species which seems so devoted to us that individuals even seem to have sacrificed for us their self, their interests, and their own feelings. Transported by men throughout the universe, submitted to all the actions that could influence their development, matched in their coupling at the will of their masters, dogs vary in color and in the abundance of their fur, which they even sometimes lose completely; in their temperament; in their size, which varies as one to five in linear dimensions or more than a hundredfold in mass; in the form of the ears, nose, and tail; in the relative length of the legs; in the progressive development of the brain in domestic varieties, which affects even the form of the head—sometimes slender, with tapering muzzle and flat forehead, sometimes with short muzzle and domed forehead: to the point that the visible differences between a mastiff and a spaniel, a greyhound and a *doguin*, are greater than those between any wild species of the same natural genus. Finally—and this is the maximum variation known to this day in the animal kingdom—there are races of dogs that have an extra digit on the rear foot, with the corresponding tarsal bones, just as in the human species there are some six-digited families. But in all these variations the relations of the bones remain the same, and the form of the teeth never changes in any appreciable manner; at the very most there are some individuals in which a false extra molar develops on one side or the other.[105]

In animals, then, there are characters that resist all influences, whether natural or human; and nothing suggests that in regard to them time has any more effect than climate.

I know that some naturalists rely a lot on the thousands of centuries that they pile up with a stroke of the pen; but in such matters we can hardly judge what a long time would produce, except by multiplying in thought

105. See my brother's memoir on the varieties of dogs, *Annals of the Museum of Natural History*, vol. 18 [1811], p. 333. This work was done at my request with skeletons I had had prepared expressly, of all the varieties of dogs. [Cuvier's younger brother Frédéric (1773–1838) was in charge of the menagerie at the Muséum.]

what a lesser time produces. Thus I have sought to collect the oldest [human] documents on the forms of animals; and there are none that, either in age or abundance, equal those furnished by Egypt. It offers us not only pictures, but the bodies of the animals themselves, embalmed in its tombs.

I have examined with the greatest care the figures of animals and birds, engraved on the numerous obelisks brought from Egypt to ancient Rome. As whole animals—which is all that their artists were able to observe—all these figures have a perfect resemblance to their originals [objets] as we see them today. My learned colleague Mr. Geoffroy Saint-Hilaire, convinced of the importance of this research, took pains to collect as many animal mummies as he was able, from the tombs and temples of upper and lower Egypt. He brought back cats, ibis, birds of prey, dogs, monkeys, crocodiles, and one ox's head, [all] embalmed; and certainly one cannot detect any greater difference between these creatures [êtres] and those we see, than between the human mummies and the skeletons of present-day men. One could find [such a difference] between the ibis mummies and the [living] ibis that naturalists have described hitherto; but I have dispelled all doubts in a memoir on this bird (which forms part of this volume), in which I have shown that it is still the same as in the time of the pharaohs. I am well aware that there I am citing monuments of only two or three thousand years ago, but that is as far back as it is possible to reach [figs. 23, 24].

Thus there is nothing in the known facts that can give the slightest support to the opinion that the new genera that I have discovered or established in the fossil state—the *palaeotheriums*, the *anoplotheriums*, the *megalonyxes*, the *mastodons*, the *pterodactyles*, etc.—could have been the root stock of any of the animals of today, differing from them only by the influence of time or climate. And [even] if it were true—which I am still far from believing—that the fossil elephants, rhinoceros, elks, and bears differ from those of the present no more than the races of dogs differ from one another, one could not conclude from that that the species were identical, because the races of dogs have been subjected to the influence of domestication, which those animals have neither undergone nor could undergo.

Moreover, when I maintain that the stony beds contain the bones of several genera, and the superficial beds those of several species, which no longer exist, I do not claim that a new creation [création] was needed to produce the existing species. I only say that they did not exist in the same places, and that they must have come there from elsewhere.

Suppose, for example, that a great irruption of the sea covered the continent of New Holland with a mass of sand or other debris. It would bury there the corpses of kangaroos, wombats, dasyures, bandicoots, flying phalangers, spiny anteaters [échidnés], and duck-billed platypuses

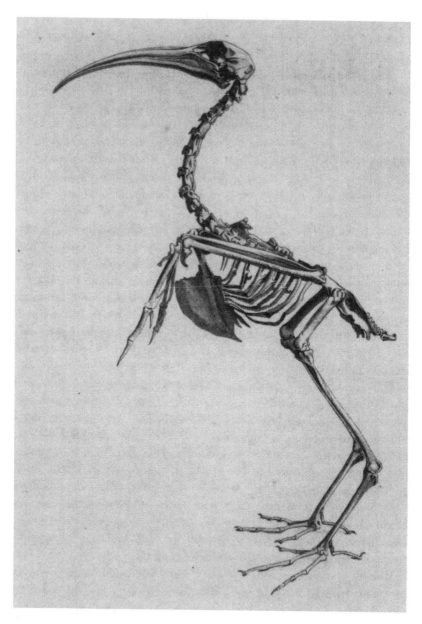

FIGURE 23 The skeleton of the sacred ibis of the ancient Egyptians, as found in mummified form in Egyptian tombs, from a paper Cuvier had published in 1804. Since it was distinctly different from the living bird commonly identified as the same species, it had been used (by other naturalists) as evidence for the gradual "transformation" or evolution of animals over long periods of time.

Numenius Ibis?,

Oiseau que je pense être le véritable Ibis des Égyptiens?

FIGURE 24 A drawing of the living bird that Cuvier regarded as the true descendant of the sacred ibis of the ancient Egyptians. He claimed that its skeleton was identical to that of mummified specimens (fig. 23), thereby rejecting any "transformist" or evolutionary interpretation. The case was so important that Cuvier gave the paper pride of place in *Ossemens fossiles,* reprinting it immediately following his "Discourse."

[ornithorhynques]; and it would entirely destroy the species of all those genera, since none of them exist in other countries. If that same revolution left high and dry the many narrow straits that separate New Holland from the continent of Asia, it would open a route for elephants, rhinoceros, buffalos, horses, camels, tigers, and all the other Asiatic animals, which would come to populate a land where they had previously been unknown. And if a naturalist, after carefully studying all that living nature, were to think of digging into the soil on which it was living, he would find the remains of wholly different beings.

What New Holland would be, in the conjecture *[supposition]* we have just made, Europe, Siberia, and a large part of America are in reality. Perhaps one day, when other countries—and New Holland itself—are examined, it will be found that they have all undergone similar revolutions, I would almost say mutual exchanges of animals *[productions]*. For let us press the conjecture further. After this migration of Asiatic animals into New Holland, suppose a second revolution were to destroy Asia, their original homeland: then one would be as hard put to know where they had come from, as one can be to find the origin of our own [animals].

[34] THERE ARE NO HUMAN FOSSIL BONES

I apply this perspective to the human species. It is certain that it has not yet been found among the fossils; and this is further proof that the fossil races [of animals] were not just varieties, since they could not have been subjected to human influence.

I say that human bones have never been found as fossils; that is, of course, among fossils properly so called. For in peats and alluvia, as in cemeteries, one can disinter human bones as well as the bones of horses and other ordinary species. But among the ancient races, among the palaeotheriums, even among the elephants and rhinoceros, not the least human bone has ever been discovered. There is scarcely a workman around Paris who does not believe that the bones with which our plaster quarries teem are in large part human bones; but as I have seen several thousands of these bones I am well qualified to state there has never been a single one belonging to our species. At Pavía I examined the sets of bones brought by Spallanzani from the island of Cerigo [Kythera]; and, notwithstanding that celebrated observer's assertion, I affirm likewise that there is none of which it can be maintained that it is human. Scheuchzer's *homo diluvii testis* [man who witnessed the Deluge] is reassigned in my fourth volume to its true genus, which is that of the *proteus* [salamander]; and in a very recent examination of it—which I made at Haarlem by the kindness of Mr. van Marum, who allowed me to uncover the parts hidden in the stone—I obtained the complete proof of what I had [already] announced.[106] Among the bones found at Canstatt there is a fragment of a human jaw and several artifacts, but it is known that the deposit was dug up without precaution, and that no note was taken of the various depths at which each specimen was found. Everywhere else the specimens taken to be human were found on examination to be of some animal, whether they had been seen in nature or only in illustrations. The truly human bones were of corpses fallen into fissures or left in old mine workings, and covered with an incrustation. It is the same with objects of human fabrication. The bits of iron found at Montmartre are the wedges that the workmen used to pack the gunpowder, and which sometimes break in the stone.[107]

However, human bones are preserved as well as those of animals, when they are in the same circumstances. In Egypt there is no difference between human mummies and those of quadrupeds. In the excavations made recently in the former church of Sainte-Geneviève, I collected human bones interred under the first ancestors, which may even belong to some princes of the line of Clovis, and which had preserved their form very well.[108] On battlefields, human skeletons are no more decomposed than those of horses, if one allows for the effects of size; and among fossils we find animals as small as the rat, still perfectly preserved.

Everything thus leads us to believe that the human species did not exist in the countries where fossil bones are found, at the time [époque] of the revolutions that buried those bones. For there would be no reason for it to have escaped wholly intact from such general catastrophes, or for its remains not to be found today like those of other animals. But I do not want to conclude that man did not exist at all before that time. They might have lived in some limited areas, from which they could have repopulated the earth after those terrible events. Perhaps also the places where they

---

106. [Martinus van Marum (1750–1837), the director of Teyler's Museum in Haarlem, had purchased this famous specimen from Scheuchzer's descendants in Switzerland. Cuvier had published his report, identifying the alleged human skeleton as that of a giant salamander, in 1809; it was during his visit to the Netherlands in 1811 that he saw the specimen for himself, and excavated it further, confirming its amphibian osteology in a staged demonstration almost as spectacular as his earlier treatment of the Parisian opossum (text 8).]

107. [Alleged human artifacts had been reported from *within* the gypsum beds quarried at Montmartre. Cuvier had to establish his claim for the very recent appearance of human beings against the contemporary opinion that they might already have been present as far back as the time of the palaeotheriums.]

108. Mr. Fourcroy has given an analysis of them. [The church—in the heart of the Latin Quarter of Paris—was secularized during the Revolution; as the "Pantheon" it became a burial place for the Republic's heroes. The "first ancestors" were presumably important burials below which the much earlier remains had been found. Clovis (c. 466–511) and his successors were regarded as the first kings of France.]

remained have been entirely ruined, and their bones buried in the depths of the present seas, with the exception of a small number of individuals who have continued the species. However it may be, the establishment of man in the countries where we have said that fossils of land animals are found (that is, in the greater part of Europe, Asia, and America) is necessarily posterior, not only to the revolutions that buried those bones, but also to those that have opened up for discovery the beds that envelop them, revolutions that are the last that the globe has undergone.[109] From this it is clear that no argument in favor of the antiquity of the human species in these various countries can be drawn, either from the bones themselves or from the more or less considerable masses of stones and earth that cover them.[110]

[35] PHYSICAL PROOFS OF THE LOW ANTIQUITY
OF THE CONTINENTS IN THEIR PRESENT STATE [111]

On the contrary, examining closely what has happened at the surface of the globe, since it was laid bare for the last time and the continents took their present form (at least in their low-lying parts), one sees clearly that this last revolution — and consequently the establishment of our present societies — cannot be very ancient. This is one of the best demonstrated and least expected results of a healthy [saine] geology, and a result that is all the more valuable in that it links natural with civil history in an uninterrupted chain.

By measuring the effects produced in a given time by agencies [causes] operating today, and by comparing them with what they have produced since they began to operate, one can work out more or less the moment at which their action started; and this is necessarily the same as that at which our continents took their present form, or that of the last sudden retreat of the waters.[112]

109. [That is, man has arrived on the scene not only since the bones were buried, but also since the bone-bearing deposits were excavated into the present valleys etc.]

110. [Cuvier originally added the following passage: "It is indeed remarkable that the oldest and most reliable witnesses *[temoignages]* of history are in accordance with natural monuments on this point: they all speak of a renewal of society that happened after a great catastrophe; and if one examines the state of society itself and the date of the memories that it has preserved, one sees that this catastrophe must be quite recent." He later deleted this, presumably when he inserted the next section.]

111. [The page numbering of the manuscript indicates that this section was written *after* all that survives of the rest of the original text. The heading, required by the sense of the text, was inserted by Cuvier at this point in the second edition of the "Discourse" (1821).]

112. [The idea of using the observable rates of present physical processes to estimate the elapsed time since the "last catastrophe" had been pioneered by Deluc; he called such processes "natural chronometers." Cuvier originally added the following passage, but deleted it before the work was printed: "Nature has also left monuments of the last catastrophe, which, without giving a wholly precise date, do nonetheless concur with human traditions in showing that it is not very ancient. We

In effect, this is to reckon that, from this retreat, our present escarpments began to crumble, and to form piles of debris at their base; that our present rivers began to flow and to deposit their alluvia; that our present vegetation began to extend and form soil; that our present cliffs began to be eaten away by the sea; that our present dunes began to be shifted by the wind: all this at the same time that human colonies began – or began again – to spread and to become established wherever nature allowed. I shall not speak at all of our volcanos, not only because of the irregularity of their eruptions, but because there is nothing to prove that they could not have existed under the sea, and thus they cannot serve to measure the time that has elapsed since its last retreat.

Mr. Deluc and Mr. Dolomieu are those who have examined most carefully the growth of detrital deposits *[atterrissemens]*. Although they are strongly opposed [to each other] on a large number of points about the theory of the earth, they are agreed about this: detrital deposits grow very quickly; they must indeed have grown even more quickly in the beginning, when the mountains supplied more material to the rivers; yet their extent is still quite limited.

Dolomieu's memoir on Egypt[113] tends to prove that in Homer's time the tongue of land on which Alexander had his town built did not yet exist; that one could navigate directly from the island of Phare into the gulf since called Lake Mareotis; and that that gulf then had the length (indicated by Menelas) of about fifteen or twenty leagues. Thus it must only have needed the nine hundred years that elapsed between Homer and Strabo to get things into the state described by the latter, and to reduce this gulf to the form of a lake six leagues in length. What is more certain is that since then things have changed still more. The sands thrown up by the sea and wind have formed, between the island of Phare and the old town, a tongue of land two hundred fathoms [400 yards] broad on which the new town was built. They have obstructed the nearest mouth of the Nile and reduced Lake Mareotis almost to nothing. During this time, the alluvia of the Nile have been deposited along the rest of the shoreline. At the time of Herodotus, the coast extended in a straight line, and still appears thus on Ptolemy's maps; but since then it has advanced and taken a semicircular form. The

---

have seen that it consisted simply in a sudden displacement of the sea, which, in abandoning the places it had previously enveloped, submerged anew most of those it had formerly left uncovered, where — at the moment of this last catastrophe — there lived men and terrestrial animals, of which almost all individuals were destroyed and their corpses covered by the waters." This indicates unambiguously that Cuvier did not regard the world before the "last catastrophe" as wholly prehuman.]

113. ["Constitution physique de l'Egypte"], *Journal of physics*, vol. 42 [1793].

towns of Rosetta and Damietta, built on the edge of the sea less than a thousand years ago, are today two leagues away.

Anyone can learn in Holland and Italy with what rapidity the beds of the Rhine, Po, and Arno—constricted today between dikes—rise in level, and how much their mouths advance into the sea, forming long promontories on the coastline; and can judge from these facts how few centuries these rivers have needed in order to deposit the low plains they now traverse. Many towns that, at times well known to history, were flourishing seaports, are now several leagues inland; several have even been ruined as a result of this change in position. Venice has scarcely maintained the lagoons that separate it from the continent; and one day, despite all its efforts, it will inevitably be joined to the mainland.[114] On Strabo's authority it is known that Ravenna was in the lagoons at the time of Augustus, as Venice is today; but at present Ravenna is a league from the shore. Spina had been founded on the coast by the Greeks, but by the time of Strabo it was ninety furlongs [about 10 miles] away, and today it has been destroyed. Adria, which gave its name to the same [Adriatic] sea of which it was the principal port some twenty centuries ago, is now six leagues away. Fortis has even made it plausible that at an earlier epoch the Euganean Hills could have been islands.[115]

My learned colleague at the Institute, Mr. de Prony, inspector general of roads and bridges, has communicated to me some very valuable information to explain these changes on the coast of the Adriatic.[116] Having been ordered by the government to examine what remedies could be introduced for the devastations caused by the floods of the Po, he established that, since the time it was enclosed by dikes, the bed of this river has risen so much that the surface of the water is now higher than the roofs of the houses in Ferrara. At the same time its deposits have advanced into the sea with such rapidity that, comparing old maps with the present state, one sees that the shore has gained more than six thousand fathoms [about 7 miles] since 1604, which makes 150 or 180 feet—and in some places 200 feet—per annum. The Adige and the Po are today higher than all the land between them; and it is only by opening new beds for them, in the

---

114. See the memoir by Mr. Forfait, on the lagoons of Venice [1800].

115. [Ravenna now lies inland from the coast of the Adriatic Sea, about seventy-five miles south of Venice; Spina (now Po di Primaro) was a town at the mouth of the Po; Adria, also on the Po delta, is about halfway between Venice and Ravenna; the Euganean Hills rise abruptly from the north Italian plain near Padua, about thirty miles west of Venice.]

116. See the note by Mr. de Prony, printed after this discourse [not included in this edition. The engineer and mathematician Gaspard Riche, baron de Prony (1755–1839), was director of the École des Ponts-et-Chaussées (School of Roads and Bridges, i.e. of civil engineering) in Paris].

lower parts that they deposited in the past, that the disasters they now threaten can be prevented.

The same causes have produced the same effects along the branches of the Rhine and the Maas; and it is thus that the richest areas of Holland have the continually frightening spectacle of rivers suspended twenty or thirty feet above the soil. Mr. Wiebeking, director of roads and bridges in the kingdom of Bavaria, wrote a memoir on this state of things—so important for both peoples and governments to understand—in which he shows that this property of raising their beds belongs more or less to all rivers.

The deposits along the coasts of the North Sea operate no less rapidly than in Italy. They can be followed easily in Friesland and in the country around Groningen, where it was known that the first dikes were constructed by the Spanish governor Gaspar Roblès in 1570. A hundred years later, three-quarters of a league of land had already been gained in some places outside these dikes; and the town of Groningen itself—built in part on ancient ground, on a limestone that does not belong at all to the present sea, in which are found the same shells as in our coarse limestone around Paris—the town of Groningen is only six leagues from the sea. Having been to these places, I can confirm on my own authority facts that are well known anyway, most of them already very well described by Mr. Deluc.[117] The same phenomenon can be observed with the same precision all along the coasts of East Friesland, in the country of Bremen and Holstein, because the time when the new lands were enclosed for the first time is known, and so one can measure how much they have gained since.

This strip of wonderful fertility, formed by the rivers and sea, is a gift for this country all the more precious in that the old soil, covered by heathland or peat, is almost everywhere resistant to agriculture. The alluvial land alone supplies subsistence to the populous towns constructed along this coast since the Middle Ages; and they could not perhaps have attained this degree of splendor without the rich lands that the rivers had prepared for them, and that they increase continually.

If the size attributed to the Sea of Azov by Herodotus, as almost equal to the Euxine [Black Sea],[118] was expressed less vaguely, and if it were known just what he meant by the Gerrhus,[119] we would still find in this some strong

117. In his *Letters to the queen of England* [i.e. his *Lettres physiques et morales* (1779). Cuvier had been there during his travels as an administrator of higher education under Napoleon, in 1810 and 1811.]

118. [*Historia,*] *Melpomene* 86. [The Sea of Azov lies to the east of the Crimea, receives the waters (and the sediment) of the Don and other rivers, and has a narrow channel connecting it with the far larger Black Sea.]

119. Ibid., 56.

proofs of the changes produced by rivers, and of their rapidity. For the alluvia of the rivers would by themselves have been able since that time—that is, in 2,250 years—to reduce the Sea of Azov to what it now is, to shut off the course of that branch of the Dnieper that would have flowed into the Hypacyris, and with it into the gulf of Carcinites or Olu-Degnitz, and even to reduce the Hypacyris and the Gerrhus to almost nothing.[120] One would have no less strong proofs if one could be certain that the Oxus or Sihoun, which now flows into the Aral Sea, once flowed into the Caspian Sea; but the witnesses on all these points are too vague, and contradict each other too much, to serve as support for physical propositions. Besides, we have near us some facts that are demonstrative enough to eliminate any ambiguity.

We have spoken above about dunes, or those hills of sand that the sea throws up on low coastlines where its bed is sandy. Wherever human industriousness has not known how to stabilize them, these dunes advance on the land as irresistibly as rivers advance on the sea. They press before them the lagoons formed by rainwater on the land they border, and which they prevent from communicating with the sea; and in many places they advance with frightening rapidity. Forests, buildings, cultivated fields: they invade all. Those of the Gulf of Gascony [Bay of Biscay] [121] have already covered a large number of villages mentioned in medieval title deeds; and at this moment, just in the department of Landes, they are menacing ten with inevitable destruction. One of these villages, that of Mimizan, has been fighting against them for fifteen years, and a dune more than sixty feet high is approaching, as it were, perceptibly. In 1802 the lagoons swamped five fine small farms in Saint-Julien; [122] long ago they covered an ancient Roman road that led from Bordeaux to Bayonne, which could still be seen thirty years ago when the water was low.[123] The Adour, which in known times passed by old Boucau and flowed into the sea at Cape Breton, is now

120. See Mr. Rennell's *Geography of Herodotus* [1800], and a part of the work of Mr. Dureau de la Malle, entitled *Physical geography of the Black Sea etc.* [1807]. N.B. Mr. Dureau (p. 170) has Herodotus make the Borysthenes [Dnieper] and the Hypanis flow into the Palus Maeotis [Sea of Azov]; but Herodotus (*Melpomene* 53) says only that these two rivers flow together into the same marsh, that is, into the Liman, as they do today. Herodotus does not make the Gerrhus and the Hypacyris go there any more. [The Dnieper is here interpreted as having once flowed, at least in part, into the Sea of Azov, east of the Crimea; it now makes a sharp westward turn and flows directly into the Black Sea, west of the Crimea.]

121. See the report on the dunes of the Gulf of Gascony by Mr. Tassin (Mont-de-Marsan, Year X [1801–2]).

122. Memoir of Mr. Brémontier, on the stabilization of dunes ["Moyens de fixer les dunes," 1797.]

123. Tassin, loc. cit.

diverted by more than a thousand fathoms [over a mile]. The late Mr. Bré-montier, inspector of roads and bridges, who carried out major works on the dunes, estimated their progress at sixty feet a year, and at certain points at seventy-two. According to those calculations they would need only two thousand years to reach Bordeaux; and given their present extent it must be a little more than four thousand years since they began to be formed.[124]

The peat bogs produced so generally in northern Europe, by the accumulation of the debris of sphagnum and other aquatic mosses, give another measure of time. They grow at a rate specific to each place; thus they envelop the little mounds of ground on which they form. Many of these mounds have been buried within human memory. In other areas the peat bogs descend the length of the valleys, advancing like glaciers; but glaciers melt at their lower edge, whereas the peat bog is stopped by nothing. By boring through it to the solid ground one can estimate its age; and one finds—for peat bogs as for dunes—that they cannot reach back to an indefinitely remote epoch. It is the same for the screes [éboulemens] that form with prodigious rapidity at the foot of all escarpments, and that are still far from having covered them. But since precise measurements have not yet been applied to these two kinds of process, we will not insist on them any further.

We see sufficiently that nature everywhere maintains the same language; that everywhere she tells us that the present order of things [l'ordre actuel des choses] does not reach back very far. And—what is indeed remarkable— mankind everywhere speaks to us like nature, whether we are studying the true traditions of peoples, or examining their moral and political state, and the intellectual development they had attained when their authentic records [monumens] begin. Let us then question the history of nations; let us read their ancient books; let us try to recognize in them that which contains real facts, and separate out the self-interested fictions that mask their truth.

[36] ALL KNOWN TRADITIONS MAKE THE RENEWAL
OF SOCIETY REACH BACK TO A MAJOR CATASTROPHE

The Pentateuch[125] has existed in its present form at least since Jeroboam's schism, for the Samaritans accept it like the Jews; that is, it is certainly

124. See Mr. Brémontier's memoir ["Moyens de fixer les dunes," 1797].

125. [Cuvier originally completed this sentence, "is the most ancient book preserved among the peoples who lived on the shores of the Mediterranean," thus indicating his primary reason for dealing with the Jewish records in first place. He then rewrote the sentence, probably because he made the same point below. The Pentateuch comprises the first five books of the Jewish scriptures, or Old Testament, and was traditionally taken to have been written by Moses himself.]

more than 2,800 years old.[126] There is no reason not to attribute the writing *[rédaction]* of Genesis to Moses himself, which would take it back another five hundred years. Moses and his people left Egypt, which all Western nations concede was the most anciently civilized kingdom of all those that surround the Mediterranean. The Jews' lawmaker had no motive for shortening the duration of the nations; and he himself would have been discredited among his own nation if he had taught a history quite contrary to what they must have learned in Egypt. Thus there is every reason to believe that, in Egypt at that time, there were no ideas about the antiquity of existing peoples other than those offered by Genesis.[127] Now, Moses recorded a general catastrophe, an irruption of the waters, an almost total regeneration of mankind; and he made its date *[époque]* reach back only fifteen or sixteen centuries before himself (according to the texts that extend that interval furthest), and consequently to at least five thousand years before us.

The same ideas seem to have reigned in Chaldea, since Berossus (who wrote in Babylon at the time of Alexander) spoke of the Deluge more or less like Moses,[128] placing it immediately before Belus, the father of Ninus. Sanchuniathon is not seen speaking of it in his history of Phoenicia, whatever may be the authenticity of this book;[129] but it appears to have been believed in Syria, for in a temple at Hierapolis there was shown—at a much later time, it is true—the fissure through which it was claimed the waters had run.[130]

As for Egypt itself, it could be believed that this tradition had been effaced there, since no explicit trace of it can any longer be found in the most ancient fragments that remain to us in that country. It is true that they are all later than the devastation of Cambyses, and indeed that the small agreement among them proves they are derived from mutilated documents. For

126. See the *Introduction to the books of the Old Testament [Einleitung in das Alte Testament]* by Eichhorn, Leipzig, 1803. [Johann Gottfried Eichhorn (1752–1827) was a leading orientalist and biblical scholar at Göttingen; his pioneer work on textual criticism reconstructed the varied sources from which the biblical texts in their present form had been edited in antiquity, and traced their subsequent history. The word "rédaction" in the next sentence is ambiguous, but it probably indicates that Cuvier accepted the biblical critics' view that Moses had assembled and edited earlier texts, rather than composing them *ab initio* (as precritical tradition assumed).]

127. [This passage shows how, in Cuvier's argument, Genesis acted simply as a kind of proxy for Egyptian records, rather than being a source of any special authority in its own right: the Egyptian civilization was known to be older than the Jewish, but its hieroglyphic records had not yet been deciphered, so that Genesis was left as the oldest record available, albeit an indirect source of evidence.]

128. Josephus, *Antiquities of the Jews*, book 1, chap. 3; Eusebius, *Preparation of the Gospel [Praeparatio evangelica]*, book 9, chap. 4; Syncellus, *Chronography*.

129. See Eusebius, *Preparation of the Gospel*, book 1, chap. 10.

130. Lucian, *On the Syrian goddess [De dea syria]*.

it is impossible to establish the smallest plausible connection between the lists of the kings of Egypt written by Herodotus under Artaxerxes, by Eratosthenes and Manethon under the Ptolemys, and by Diodorus under Augustus; even the different extracts drawn from Manethon do not agree with each other.[131] Failing history, however, the mythology of Egypt does seem to recall these great events, in the adventures of Typhon and Osiris; and if the priests of Saïs really told Solon the stories that—following him—Critias reports in Plato, one would even have to believe they had preserved more precise notions of a major revolution, although they made it date back to a time long before Moses. They had even established in theory an alternation of revolutions, some operated by water, the others by fire; an idea that was also spread among the Assyrians, and as far as Etruria.

The Greeks, among whom civilization arrived from Phoenicia and Egypt, and so late, mixed Phoenician and Egyptian mythologies—of which they had been brought confused notions—with the no less confused traits of their own early history. The personified sun, named Ammon (or the Jupiter of Egypt), became a Cretan prince; Ptah, the maker of all things, was Hephaestus or Vulcan, a blacksmith of Lemnos; Chou, another symbol of the sun or of the divine force, was transformed into a strong Theban hero, their Heracles or Hercules; the cruel Moloch of the Phoenicians, the Remphah of the Egyptians, was the Chronos or Time who devours his children, and then Saturn, king of Italy.[132] If a somewhat violent inundation occurred, under one of their princes, they described it subsequently with all the circumstances vaguely remaining in their memory of the great cataclysm;[133] and they had the earth repeopled by Deucalion, all the while leaving a long posterity to his uncle Atlas.

However, the incoherence of [all] these narratives, which attests to the barbarism and ignorance of all the peoples on the shores of the Mediterranean, indicates equally the recentness of their establishment; and that recentness is itself strong evidence of a major catastrophe. In Egypt there is indeed talk of hundreds of centuries; but they are filled with gods and demigods. Today it has been proved, so to speak, that in the sequence of years and human kings, placed after the demigods and before the invasion

---

131. See the English *Universal history,* [i.e. ed. George Sale et al.], vol. 1 [1780].

132. See Jablonski, *Pantheon of the Egyptians [Pantheon Aegyptiorum],* and Gatterer's paper *Origin of the Egyptian gods [De theogonia Aegyptiorum]* in the Göttingen memoirs, vol. 7 [1786]. These two authors do not agree—any more than the ancients—about the meaning of the Egyptian divinities; but they do agree—with all those same ancients—on the gross alterations that the Greeks imposed on them.

133. [Cuvier first used the conventional term "universal deluge," but then presumably remembered that his own conception of the event was that it had *not* been universal, but confined to low-lying areas.]

of the pastoral people, what have been taken to be successive kings should be interpreted as the chieftains of several contemporary little states.[134] Macrobius[135] is sure that there were records of observations of eclipses made in Egypt, that would imply work continued without interruption since at least twelve hundred years before Alexander. But how did Ptolemy not deign to use any of these observations, made in the country where he was writing?

There was still no great empire in Asia at the time of Moses; and the Greeks themselves, despite their facility at inventing fables, did not take the trouble to fabricate a [high] antiquity. The oldest colonists of Egypt or Phoenicia, who came to snatch them from a state of savagery, do not reach back more than four thousand years before the present time; and the oldest authors who speak of them do not date to three thousand. The Phoenicians themselves were in Syria only a short time, when they made colonies in Greece. The astronomical observations of the Chaldeans, transmitted to Aristotle by Callisthenes, would also reach back four thousand years, if that fact, reported only by Simplicius (six hundred years after Aristotle) is at all authentic; which is very doubtful, since the Chaldean observations of eclipses, actually preserved and cited by Ptolemy, go back only to twenty-five hundred years. However that may be, the empire of Babylon or that of Assyria could not have been powerful, or have left unsubjected around them such lesser tribes as all those of Syria, long before what is called the Second Kingdom of Assyria. The thousands of years that the Chaldeans attributed to themselves are thus as fabulous as those of the Egyptians; or rather, they are only astronomical periods calculated retrospectively from inexact observations, or even simple cycles chosen arbitrarily and multiplied by themselves.[136]

The most reasonable of the ancients had no other ideas,[137] and did not put the earliest of the conquerors, their Ninus and Semiramis, back more than about forty centuries, after which history maintained a long silence; which makes one suspect that they could indeed be nothing but later creations by historians.

Our present knowledge and civilization descends without interruption from the Egyptians and Phoenicians, by way of the Greeks and Romans; the

---

134. Gatterer, [Origin of the Egyptian gods], and Marsham's System [i.e. probably his Chronicus, 1672]. [In place of this last sentence, Cuvier first wrote, "Perhaps the history of the first human kings was even fabricated from poorly understood hieroglyphs."]

135. Scipio's dream [Somnium Scipionis], 21.

136. See Mr. de Guignes's memoir on the sares of the Babylonians, Academy of Belles-Lettres, vol. 47 [1809]; and Mr. Gentil's Voyage [to the Indies], [1779–81], vol. 1, p. 241.

137. See Velleius Paterculus [Historiae Romanae] and Justin [Historiae Philippicae].

Jews have given us directly our purest ideas of morality and religion;[138] some enlightened strands have also come to us through them and the Greeks, Chaldeans, Persians, and Indians. What is remarkable is that these peoples form a single race; they resemble each other in facial appearance, and indeed by an infinity of conventions such as their divinities and the names of their constellations, and lastly even by the roots of their languages.[139]

Among these peoples, those whose civilization is perhaps the most ancient and seems to have varied least in its forms—those who are still probably closest to its cradle—the Indians unfortunately have no history at all. Among the infinity of books on mystical theology and abstruse metaphysics that we possess, there are none that can tell us systematically about their origin and the vicissitudes of their societies. Their *Maha-Bharata*, a so-called great history, is only a poem; their *Pouranas* are only legends; and in comparing them with Greek and Roman authors it is only with great effort that some scraps of a kind of incomplete chronology have been established, interrupted at each instant, which reaches no further back than Alexander.[140] Today it has been proved that their astronomical tables, from which they also wanted to deduce their extreme antiquity, were calculated in retrospect;[141] and it has just been recognized that their *Surya-Siddhanta*, which they regard as their oldest scientific treatise on astronomy, and which they claim has been revealed for more than two million years, cannot have been composed more than about 750 years ago.[142] Their sacred books, or *Vedas*, may date back 3,200 years—which would be near the time of Moses—judging from the calendar that is attached to them and to which they relate, and from the position of the colures that that calendar indicates.[143]

138. [Cuvier first wrote simply, "The Jews added to it profoundly"; the change conceded prudently that the Jewish scriptures were not merely derivative, at least in the sphere of morals and religion.]

139. On the analogy between the languages of India, Persia, and the West, see Adelung's *Mithridates* [General linguistics, 1806–17]; on the analogy of the divinities of the Indians, Egyptians, Greeks, and Romans, the works by Jablonski and Gatterer cited above [note 132], and William Jones's memoir (with notes by Mr. Langlès) [on gods of Greece, Italy, and India] in the first volume of the French translation [1805] of the Calcutta memoirs, p. 162ff. The identity of the constellations, and principally the signs of the zodiac of the Indians and the peoples farther west, those of the names of the days of the week, etc., are now known to everyone.

140. See the great work of Mr. Paterson [in fact by Wilford], on the chronology of the kings of Magadha, emperors of India, and on the epochs of Vicramadityia and Salahanna, in *Calcutta memoirs*, vol. 9 [1807].

141. See Mr. de Laplace, *Explanation of the world system* [*Exposition du système du monde*, 1798–99], p. 330.

142. See the paper by Mr. Bentley on the antiquity of the *Surya-Siddhanta*, *Calcutta memoirs*, vol. 6, p. 537 [1799]; and the paper by the same author, on the Indians' astronomical systems, ibid., vol. 9, p. 195 [1805].

143. See the paper by Mr. Colebrooke on the Vedas, and particularly p. 493, in vol. 8 of the *Calcutta memoirs* [1805]. [The colures are great circles passing through the celestial poles, dividing the zodiac into two groups of six signs each; their changing positions record the slow precession of the equinoxes.]

However, the Indians have not totally forgotten the revolutions of the globe: their theology mentions the successive destructions that its surface has already suffered and that it must yet suffer; and they date the last to only a little less than five thousand years.[144] One of these revolutions is even described in terms that almost correspond to those of Moses.[145] What is no less remarkable is that the time at which they place the start of their human sovereigns (those of the race of Sun and Moon) is almost the same as that at which those of the Assyrians start, about four thousand years before the present time.

It is useless to ask about these great events among the more southerly people such as the Arabs and Abyssinians: their ancient books no longer exist. They have no history apart from what they have made for themselves recently, and which they have modeled on the Bible: thus what they say of the Deluge is borrowed from Genesis, and adds nothing to its authority. But the Mazdeans *[Guèbres]*, today the sole repository of the teaching of Zoroaster and the ancient Persians, also place a universal deluge before Cayoumarats, whom they make their first king.

To recover truly historical traces of the last catastrophe, it is necessary to go beyond the great deserts of Tartary. To the east and north lives another race, all of whose institutions and practices differ from ours as much as their appearance and temperament. They speak in monosyllables; they write in arbitrary hieroglyphs; they have a moral politics only, without religion (for the superstitions of Fo [i.e. Buddhism] have reached them from India). Their yellow color, prominent cheeks, narrow slanting eyes, and thin beard make them so different from us that it is tempting to believe that their ancestors and ours escaped from the great catastrophe on two different shores; but however that may be, they date their deluge from more or less the same epoch as we do.

The *Chou-king* [Shujing] is the oldest book of the Chinese;[146] it is said to have been edited by Confucius with fragments of earlier works, about 2,250 years ago. Two hundred years later, under the emperor Chi-hoang-ti [Shi Huangdi], there was the persecution of scholars and destruction of books. One part of the *Chou-king* was reconstructed from memory by an aged scholar forty years later, and another was recovered from a tomb; but almost half was lost forever. Now this book, the most authentic in China, begins the history of that country with an emperor named Yao. He is de-

---

144. Le Gentil, *Voyage to the Indies* [1779–81], p. 235; Bentley, *Calcutta memoirs* vol. 8, p. 222; Paterson [i.e. Wilford], ibid., p. 86.

145. William Jones, *Calcutta memoirs*, French translation [ed. Labaume], vol. 1, p. 170.

146. See the preface to the edition of the *Chou-king*, given by Mr. de Guignes [1770]. [The *Shujing*, or "Book of history" (or documents), is one of the Confucian classics.]

picted as being concerned with getting waters drained away, "which, being elevated to the heavens, still bathed the feet of the highest mountains, covered the lower hills, and made the plains impassible." This Yao dates from 4,500 years before the present, according to some, or 3,930 years according to others; the variety of opinions on this date even extends to 284 years.

Some pages further on, we are shown Yu, a minister and engineer, reestablishing the watercourses, raising dikes, digging canals, and regulating the taxes for each province in the whole of China, that is, in an empire of six hundred leagues in every direction; but the impossibility of such operations—after such events—clearly shows that this is only a moral and political tale [roman]. More recent historians have added a series of emperors before Yao, but with a mass of fabulous circumstances, without daring to assign them fixed dates, varying endlessly among themselves (even on their number and their names), and without being approved by all their compatriots.[147]

It is to Yao that the introduction of astronomy into China is attributed; but the true eclipses recorded by Confucius in his chronicle of the Kingdom of Lou [Lu] go back only 2,600 years, barely half a century further than that of the Chaldeans recorded by Ptolemy.[148] One is indeed found in the *Chou-king*, dating to 3,965 years, but it is recorded with such absurd circumstances that it is probable that the account was added after the event. A conjunction at 4,259 years, which would be the oldest known observation, is still controversial. The first that appears reliable is an observation of Gnomon, at 2,900 years.

Is it possible that it is simply chance that gives such a striking result, and that has the traditional origins of the Assyrian, Indian, and Chinese monarchies dating back more or less forty centuries? Would the ideas of peoples who have had so little connection with each other—whose language, religion, and laws have nothing in common—be in accord on this point, unless they were based on the truth?

We shall not ask for precise dates from the Americans, who had no true writing, and whose oldest traditions go back only a few centuries before the arrival of the Spaniards; nonetheless, some traces of a deluge are believed to be perceptible in their crude hieroglyphs.[149]

147. [Shu Huangdi (d. 210 B.C.) was the first emperor of the Qin dynasty. Yao and Yu are legendary superheroes from the third millennium B.C.]

148. [Lu was the home "state" of Confucius. The manuscript text of the "Discourse" does not survive beyond this sentence, so that the published text cannot be further compared with Cuvier's original draft.]

149. See the excellent and magnificent work of Mr. de Humboldt, on the Mexican monuments [*Vues des Cordillères*, 1810].

The most degraded of human races, that of the negroes, whose form approaches most closely to that of the beast, and whose intelligence has nowhere risen to the point of reaching a regular form of government or the least appearance of sustained knowledge, has nowhere preserved either annals or traditions. Thus it cannot inform us on what we seek, although all its characters show clearly that it escaped from the great catastrophe at another point than the Caucasian and Altaic races, from which it had perhaps been separated long before that catastrophe took place.[150]

Thus all the nations that can speak to us testify that they have been renewed recently, after a great revolution of nature.

[37] THE ASTRONOMICAL MONUMENTS LEFT BY THE ANCIENTS CANNOT
BEAR THE EXCESSIVELY REMOTE DATES THAT HAVE BEEN CLAIMED [151]

This unanimity of historical or traditional witnesses, on the recent renewal of mankind, and their accord with those drawn from the operations of nature, would doubtless excuse us from examining some ambiguous monuments, of which some persons want to avail themselves in order to uphold a contrary opinion. But that very examination—to judge from some attempts—would probably only add further proofs to what traditions tell us.

Today it appears that the famous zodiac on the doorway of the temple of Dendera cannot sustain it [i.e. an extreme antiquity]: for nothing proves that its division into two bands, each of six signs, indicates the position of colures resulting from the precession of the equinoxes, or corresponds simply to the start of the civil year at the time it was designed; a year that, being in Egypt of only 365 days exactly, made the rounds of the zodiac in 1,508 years, or, according to what the Egyptians imagined, in 1,460 years (which proves that they did not observe effectively). A fact that makes this supposition plausible is that inside the same temple there is another zodiac, in which it is the Virgin [Virgo] that starts the year. If it were a matter of the position of the solstice, the zodiac inside would have been made two thousand years before that on the portico; but conceding on the contrary that they wanted to indicate the start of the civil year, an interval of a little over a hundred years would suffice.

150. [Cuvier's opinions, however distasteful to modern sensibilities, are those of his time and place. It should be noted that his conjecture that the three main races had survived the "last catastrophe" at different points on the earth's surface did not assume the intrinsic superiority of the Caucasians; the same conjecture, of course, reduced the version of events given in Genesis to the status of one local account.]

151. [The previous heading is the last to be printed in the margin of the original edition of the "Discourse." This and all subsequent headings appear in the equivalent positions in the second edition (1821), and are required by the sense of the text.]

It remains to be known whether our zodiac does not contain in itself some proofs of its antiquity, and whether the figures that have been given to the constellations had any connection with the position of the colures at the time they were conceived. Now all that has been said in this respect is based on the allegories that have been claimed to be seen in these figures: that the Balance [Libra], for example, indicates the equality of days and nights; the Bull [Aries], plowing; the Crab [Cancer], the turning of the sun; the Virgin [Virgo], the harvest, and so on; and how much of all that is by chance? Moreover, these explanations ought to vary for each country, such that the zodiac would have to be given a different date, according to the climate to which its invention was assigned; indeed, perhaps there is no climate and no time at which a natural explanation could be found for all the signs. Lastly, who knows whether the names were not given very long ago, in an abstract manner, to the divisions of space or time, or to the sun in its different states, just as astronomers now give them to what they call signs; or whether they were not applied to constellations of groups of stars, at a time fixed by chance, such that nothing could be concluded from their signification.[152]

But—it will be said—is not the state in which we find astronomy among the ancient peoples a proof of their antiquity? And did not the Chaldeans and Indians need many centuries of observations, in order to attain the knowledge that they already had three thousand years ago, on the length of the year, the precession of the equinoxes, the relative movements of the moon and sun, etc.? But has anyone calculated the progress that a science made in a nation that had no other science, and in which the serenity of the sky, a pastoral life, and superstition made the stars the object of general contemplation, and in which colleges of the most respected men were charged with observing them and recording their observations in writing? If among these many individuals, who had nothing else to do, there were found one or two geometers, all that these peoples knew could have been discovered in a few centuries.

Let us consider that real astronomy has had only two [creative] periods since the Chaldeans: that of the school of Alexandria, which lasted four hundred years; and ours, which has not been as long. The period of the Arabs scarcely added anything, and all the other centuries have been as nothing to it. There have not been three hundred years between Copernicus and the author [Laplace] of *Celestial mechanics*, and [yet] it is claimed

152. See the paper by Mr. de Guignes on the zodiacs of the Orientals, *Academy of Belles-Lettres*, vol. 47 [1809].

that the Indians would have needed thousands of years to find their [astro-nomical] rules.

Moreover, even if all that has been imagined about the antiquity of as-tronomy were to be as well proved as it appears to us to be devoid of proof, nothing could be concluded from it against the great catastrophe, of which there remain to us documents that are demonstrative in quite other ways. It would only be necessary to concede, with some modern [writers], that as-tronomy was among the bodies of knowledge preserved by the men spared by that catastrophe.

[38] FALSE CONCLUSIONS ABOUT CERTAIN MINE WORKINGS

The antiquity of certain mine workings has also been much exaggerated. One very recent author claimed that the mines on the island of Elba, judg-ing by their tip heaps [déblais], must have been exploited for more than forty thousand years; but another author who also examined the debris with care reduces that period to little more than five thousand years,[153] while still supposing that the ancients exploited annually only a quarter of what is exploited now. But what reason is there to believe that the Romans, for example, extracted so little from these mines, since they used so much iron for their armies? Moreover, if these mines had been worked even for four thousand years, why was iron so little known in early antiquity?

[39] GENERAL CONCLUSION ON THE TIME OF THE LAST REVOLUTION

Like Mr. Deluc and Mr. Dolomieu, I think therefore that if there is any-thing established in geology it is that the surface of our globe has been the victim of a great and sudden revolution, the date of which cannot reach back much more than five or six thousand years; that in this revolution the countries in which men and the species of animals now best known previ-ously lived, sank and disappeared; that conversely it laid dry the bed of the previous sea, and made it into the countries that are now inhabited; that since that revolution the small number of individuals spared by it have spread out and reproduced on the land newly laid dry; and that conse-quently it is only since that time that our societies have resumed a progres-sive course, that they have formed institutions, erected monuments, col-lected facts of nature, and combined them into scientific systems.

But these countries that are inhabited today, which the last revolution laid dry, had already been inhabited previously, if not by men then at least by terrestrial animals. Consequently one previous revolution, at least, had

153. See Mr. de Fortia d'Urban, *History of China before the deluge of Ogyges* [1807], vol. 2, p. 33.

already put them under water; and judging by the different kinds of animals of which the remains are found, they had perhaps suffered up to two or three invasions by the sea.

These alternations now appear to me to be the most important geological problem to be resolved, or rather, to be clearly defined and circumscribed; for to resolve it entirely it would be necessary to discover the cause of these events, an enterprise of quite another difficulty.

I repeat: we see fairly clearly what is going on at the surface of the continents in their present state. We have grasped fairly well the uniform course and regular succession of the Primitive formations, but the study of the Secondary formations has scarcely been sketched out. That marvelous series of unknown zoophytes and marine mollusks, followed by equally unknown reptiles and freshwater fish, replaced in turn by other zoophytes and mollusks closer to those of today; those land animals and mollusks, and other freshwater animals still unknown, which then come to occupy their places, only to be chased out again, but by mollusks and other animals similar to those of our seas; the relations between these varied organisms and the plants whose debris accompanies theirs, and the relation between those two kingdoms and the mineral beds that conceal them; the lack of uniformity of the one and the other in different basins: there [indeed] is a set of phenomena that now, it seems to me, calls imperiously for the attention of [natural] philosophers.

This study, which is interesting in the variety of the products of the local or general revolutions of that time, and in the abundance of the diverse species that figure alternately on the scene, has none of the aridity of that of the Primordial formations; and unlike the latter it does not almost inevitably involve hypotheses. The facts are so pressing, so curious, and so obvious that they satisfy, as it were, the most ardent imagination; and the conclusions to which from time to time they lead, however cautious the observer, having nothing vague about them, have nothing arbitrary about them either. Finally, it is in these events closest to us that we can hope to find some traces of more ancient events and of their causes, if after so many attempts we may still flatter ourselves with such a hope.

These ideas have pursued me—I could almost say, tormented me—while I carried out the research on fossil bones that I now present to the public in collected form; research that encompasses such a small part of those phenomena of the earth's penultimate age, but which is connected in an intimate way to all the others. It was almost impossible for it not to awaken the desire to study the generality of these phenomena, at least in a limited space around us [in Paris]. My excellent friend Mr. Brongniart, to whom other studies had given the same desire, was willing to collaborate with me, and it is thus that we laid the first foundations for our work on the environs of Paris. But although that work still bears my name, it has become almost entirely that of my friend, as a result of the infinite care he has given—ever since the conception of our original plan, and our travels [in the field]—to the profound study of the objects, and to the drafting of it all. With the consent of Mr. Brongniart, I am attaching it to the present discourse, of which it seems to me capable of being an integral part, and of which it is certainly the best proof.[154]

We see in it the history of the most recent changes that have taken place in one particular basin; and it takes us as far [back] as the chalk, the extent of which on the globe is infinitely greater than that of the materials of the Paris Basin. The chalk, which had been thought so modern, thus finds itself pushed far back into the centuries of the penultimate age. It would now be important to examine the other basins that the chalk may enclose, and in general all the beds that overlie it, in order to compare them with those around Paris. The chalk itself perhaps offers some succession of organisms. It is surrounded and supported by the compact limestone that occupies the greater part of France and Germany, and of which the fossils differ infinitely from all those of our [Paris] basin. But in following it from the chalk to the almost shell-less limestone of the central ridges of the Jura, or to the conglomerates *[aggrégats]* on the slopes of the Harz, Vosges, and Black Forest, would not many more variations be found? Are not the gryphites, ammonites, and entrochites with which they teem distributed by genera, or at least by species?[155]

That compact limestone is not covered everywhere by chalk; in several places it surrounds basins [of Tertiary formations] without that intermediary, or underlies plateaus that are no less worthy of attention than those that have the chalk as their limit. Who for example will give us a history of the plaster quarries of Aix[-en-Provence], where—as in those of Paris—reptiles and freshwater fish are found, and probably also terrestrial quad-

154. [In the first volume of *Ossemens fossiles* the "Discourse" was bound with the full version of the joint monograph, with only the brief paper on the ibis between them.]

155. [The limestones and conglomerates were the Secondary formations below the Chalk (in modern terms, largely of Jurassic and Triassic age), clearly underlain in turn by the still older rocks of the Harz, Vosges, and Black Forest massifs; the limestones contained fossils quite distinct from those of the Parisian formations ("gryphites" are the oysterlike shells of *Gryphaea;* "entrochites" are crinoid ossicles). Cuvier's point is that more detailed study of these fossils might reveal much finer distinctions between the specific formations.]

rupeds, while there is nothing similar in almost two hundred leagues of in-
tervening country?

It would also be very important to know about the long series of sandy
hills that rest on both flanks of the Apennines, almost the whole length of
Italy, and which everywhere contain perfectly preserved shells (often still
with their color and mother-of-pearl), several of which resemble those of
our seas. It would be necessary to follow all the beds, to identify the fossils
in each, and to compare them with those of other recent beds, for example
those of our [Paris] environs; to connect the sequence on the one hand
with the older and more solid formations, and on the other with the recent
alluvium of the Po, the Arno, and their tributaries; to fix their relations
with the innumerable masses of volcanic products that are interposed between them; and finally to examine the relative situation of the various
kinds of shells, and of the bones of elephants, rhinoceros, hippopotamuses,
whales, sperm whales, and dolphins in which many of these hills abound.
About these low hills of the Apennines I have only the superficial knowledge
that I could get on a journey made for other reasons; [156] but I am persuaded
that they conceal the true secret of the last operations of the sea.

How many other beds are there—even celebrated for their fossils—
which we do not yet know how to tie into the general series, and of which
the relative age is consequently still indeterminate? The coppery shales of
Thuringia are said to be full of freshwater fish, and exceeding most of the
Secondary formations in age. But what is the true position of the fetid
shales of Oeningen that are also said to be full of freshwater fish; of those of
Verona, evidently full of marine fish, but fish very poorly named by the naturalists that have described them; of the black shales of Glarus and the
white shales of Eichstatt, again full of fish, lobsters, and other marine animals other than shells? [157] I find no distinct response to these questions in
our geologists' books. And we are not told why shells are found everywhere,
and fish only in a small number of places.

It seems to me that a sustained history of such singular deposits would
be worth far more than contradictory conjectures of the first origin of the
planets, and on phenomena that it is conceded cannot resemble in any way

156. [In 1809, to organize higher education in the newly enlarged French empire.]

157. [In modern terms these famous fossil localities are, respectively, a formation (Kupferschiefe)
of Permian age in Thuringia; a Swiss deposit of Miocene age; an Eocene limestone in northern Italy
(better known as that of Monte Bolca); a shale of Cretaceous age in the Swiss Alps; and a Jurassic
limestone in Bavaria (better known as the "lithographic stone" of Solnhofen). Cuvier's listing of
them—random in terms of modern knowledge—indicates the lack of just the kind of stratigraphical research he was urging here.]

those of our present physical world, and that consequently find neither materials nor touchstone there. Many of our geologists resemble those historians who—in the history of France—are only interested in what happened before Julius Caesar; their imagination has to supplement the monuments, and each of them writes a novel *[roman]* to his own taste. What would happen if these historians were not aided in their syntheses by a knowledge of later facts? Now, our geologists neglect precisely these later facts, which could at least reflect some faint light toward the night of preceding times. How good it would be, however, to have the organic productions of nature in their chronological order, as we have the main mineral substances![158] The science of [living] organization itself would gain from it; the developments of life, the succession of its forms, the precise identification of those that appeared first, the simultaneous birth of certain species, and their gradual destruction, would perhaps tell us as much about the essence of the organism, as all the experiments that we can attempt on living species. And man, to whom has been accorded only an instant on earth, would have the glory of reconstructing *[refaire]* the history of the thousands of centuries that preceded his existence, and of the thousands of beings that have not been his contemporaries!

Translated from "Discours préliminaire" in Cuvier, Ossemens fossiles *(1812), vol 1. In the original text, unnumbered section headings are printed in the margins; here they are given numbers for easy reference, and are used to break up the text; a few extra headings have been added from the second edition (the numbering is not the same as that used in Jameson's English editions). Some of the more important changes that Cuvier made in manuscript, or while the text was in press, are quoted in the footnotes (the manuscript, of which the final sections are unfortunately lost, is in MS 631, Bibliothèque Centrale, Muséum National d'Histoire Naturelle, Paris).*

158. [That is, the general outline of the order of the formations—as recognized by their rock types—was becoming clear, but not the order of the fossils that the Secondaries contained.]

16

CONCLUSIONS

––––––––––––

This book has been designed to make Cuvier's main geological writings accessible to English-speaking readers, not to describe or assess their reception, let alone to offer a biography. Cuvier's later work, and the further history of his *Ossemens fossiles,* can therefore be summarized very briefly.

Once the *Ossemens fossiles* was completed, Cuvier immediately turned to the publication of his other magnum opus, his study of the comparative anatomy and classification of the whole animal kingdom (*Règne animal,* 1817). In fact, a brief outline of his radically new "map" of the animal kingdom was published in the *Annales du Muséum* in the same year (1812) as the *Ossemens fossiles.* In place of the traditional dichotomy between animals with and without a backbone, Cuvier proposed a fourfold division that undercut the basis for any linear "scale of beings," and therefore also for any simple transformist or evolutionary explanation of the diversity of organisms. The "Vertebrata" were demoted to become just one of four radically distinct "branches" *[embranchements];* or, to put it another way, the invertebrates were split into three great branches, as distinct from one another as each was from the vertebrates. The "Mollusca" contained much the same range of animals as in a modern definition, most of them with external shells. The "Articulata" contained a wide range of segmented animals, including all the arthropods as now defined, together with "worms" of many kinds. The "Radiata" were more of a ragbag, but contained all the echinoderms and coelenterates as now defined, together with many other relatively simple invertebrates. Cuvier's branches may have been crude by the standards of the phyla of modern classifications. But they marked a decisive recognition that the diversity of animals cannot be represented on any linear scale, and that the invertebrates include several distinct groups with radically different kinds of anatomy. In its own field, Cuvier's mapping of the *Règne animal* was as important and as influential as his *Ossemens fossiles,* throughout the nineteenth century and beyond.

The "Preliminary discourse" of *Ossemens fossiles* was translated almost immediately into English, in an edition by Robert Jameson, the professor of natural history at Edinburgh. His preface and editorial notes set the tone for the reception of Cuvier's work in the anglophone world, for Jameson maintained that its main purpose was to demonstrate the historicity of the Deluge and hence to vindicate the authority of the Bible. Jameson also chose a title about which Cuvier must have felt—to say the least—highly ambivalent: English-speaking readers were presented with an *Essay on the theory of the earth* (1813). Even if there had not been a major war in progress between Cuvier's nation and Jameson's, Cuvier would have had little or no control over such an edition, in the absence of international copyright agreements. Since it was beyond his control, he seems to have withheld comment, and there is no evidence to show what he thought of it. In any event, Jameson's book was highly successful in Britain, and three editions were published, progressively amplified by Jameson's comments, even before Cuvier published a second and much revised edition of the *Ossemens fossiles.*

English-speaking readers could thus learn about Cuvier's geological conclusions—albeit with a pronounced editorial slant—in their own language, almost as soon as they were published. The bulk of Cuvier's work, however, remained accessible only to those who could obtain and read the original. In the German-speaking world, by contrast, it was Cuvier's and Brongniart's detailed work on the geology of the Paris region that was quickly appreciated and translated, at least in a full summary, in one of the leading scientific journals (*Annalen der Physik,* 1813); only later (1816) did the same journal publish a paraphrase of the "Discourse."

In 1821 Cuvier began to publish a second edition of *Ossemens fossiles.* It was bound in seven volumes instead of four, which reflected enlargements in almost every part. The publication of the first edition had brought Cuvier a further influx of new material on specific fossil animals, so that many of the specialized papers were much fuller and more conclusive. To the range of extinct animals described, perhaps the most significant additions—for the most part discovered since the first edition—were the ichthyosaur and the plesiosaur. These strange marine reptiles from the Secondary formations (in modern terms, from the Jurassic) confirmed Cuvier's earlier hunch that an age of reptiles had preceded that of the mammals; but here, in effect, Cuvier was merely giving the stamp of his authority to work that had mainly been done by others. That reflected a general change in the character of his own work, as he shifted from being a bold innovator to being mainly an authoritative synthesizer. Likewise the enlargement of his and Brongniart's geological monograph—now occupying one whole volume—was entirely due to Brongniart's greatly extended fieldwork; it had grown into an authoritative study of the "Tertiary" formations (as the younger Secondaries were now called) throughout Europe, but Cuvier himself had had no further part in it.

The "Discourse" too underwent enlargement, but without major change of content. That in itself was a sign of Cuvier's virtual withdrawal from the field. The preceding years had been rich in publications that, in effect, pursued the research agenda he had suggested, yet that newer research was barely mentioned, and certainly did not affect Cuvier's conclusions. In fact most of the enlargement of the "Discourse" was at one, quite different, point: Cuvier's discussion of the historical evidence for the antiquity of human civilizations (text 19, sec. 36) was enlarged to about four times its original length. This greatly accentuated what was, even in the first edition, a surprising feature of the "Discourse," namely the space given to the evaluation of purely textual rather than natural evidence for the past. To say the change reflected Cuvier's increasing interest in literary culture may be to put the cart before the horse.[1] That shift of interest may have been due to his perception of the growing importance of the textual evidence for refuting claims that human civilizations were immensely ancient, for such claims threatened his own inferences about a relatively recent catastrophe that had wiped out "his" fauna of spectacular mammals and brought to an end the virtually prehuman world he had claimed to reconstruct.

Only when he issued a third (and almost unchanged) edition of *Ossemens fossiles* (1825), soon after the completion of the second, did Cuvier at last sanction the publication of the "Discourse" as a separate small volume. By that time there had been a complete translation into German

---

1. His literary ambitions were crowned late in his life, when he was elected a member of the Académie des Inscriptions et Belles-Lettres; he was one of very few savants, then or since, to belong both to this prestigious literary body and to its scientific counterpart, the Académie Royale des Sciences (the successor, after the restoration of the monarchy, of the First Class of the Institut).

FIGURE 25 Cuvier as a grandee of science: the portrait printed in the first separate French edition (1826) of his "Discourse." It is a lithograph after a painting by Nicolas Jacques.

(*Ansichten von der Urwelt*, 1822) and no fewer than four editions of Jameson's *Essay* in English. Probably Cuvier reckoned that by this time an edition in French would no longer harm the sales of the larger work; certainly its publication marked his rising fame among the literate public, far beyond the circles of his fellow savants (fig. 25). In any case he chose a title making no reference to the "theory of the earth," a genre toward which— as we have seen—his attitude had been extremely cautious, when not highly critical. Instead, he entitled his work a *Discours sur les révolutions de la surface du globe* (1826).

The term "revolutions" was at first sight a striking one to use in the political climate of the Restoration, which had in effect tried to put the clock back, after the fall of Napoleon, to the monarchical regime before *the* Revolution. But as we have seen, the word had a long history as a general term for *any* major changes, whether sudden or not. In any case, Cuvier's vision of earth history was one in which the more catastrophic "revolutions" of the prehuman world had happily been replaced by the orderly calm of the present world. The prospect of a possible *future* revolution, fleetingly expressed in one of his earliest articles (text 5), had long since vanished. So the "Discourse" could be—and probably was— read as a safely conservative image of the natural world as a mirror of the political: in both, the turmoil of *violent* "revolutions" was now a thing of the past.

By the time Cuvier died in 1832, his "Discourse" had reached its sixth edition in French (counting the two embodied in the *Ossemens fossiles*) and Jameson's *Essay* its fifth. With an Italian edition (1828) added to the earlier German one, and with Jameson's republished in the United States, Cuvier's geological conclusions could and did reach educated readers virtually everywhere in the Western world.

Meanwhile, however, he himself had shifted most of his attention elsewhere. Even in the early 1820s the revision of *Ossemens fossiles* had occupied only a small fraction of his time. His strictly zoological work continued, with the revision of his *Règne animal* (2nd ed., 1829–30), and the start of a major work on the comparative anatomy of the fishes. But above all, his time was increasingly occupied by ever-growing official duties: as permanent secretary of the Académie Royale des Sciences (which had replaced the scientific Class of the Institut); as a high-level educational administrator; and, in the last years of his life, as the top administrator of the links between the government and the Protestant churches, which remained Cuvier's own cultural group within French society.

Cuvier and his geology present several apparent paradoxes. He is widely supposed to have designed his theories to support the authority of the Bible, yet his writings show a critical skepticism about the reliability of *all* ancient texts, a lack of special pleading on behalf of Genesis, and a total abstinence from the natural theology that was so common among his anglophone contemporaries. He is still often reviled as the arch-enemy of all theories of organic evolution, and he was indeed adamantly opposed to Lamarck's transformism; but his writings are more striking for their extreme caution about speculating on the origins of species or larger groups, and he never argued for their supernatural creation. He is supposed to have held back the progress of the earth sciences by advocating an extreme "catastrophism" that ignored the power of "actual causes" observable in the present; his writings, however, show that both his fossil anatomy and his geology were self-consciously based on careful "actualistic" comparison with living animals and present geological processes, and that he invoked catastrophes only where, in his opinion, present processes were clearly inadequate to explain what could be observed. Above all, his later reputation was as a highly speculative "theorist of the earth," yet in his writing he repeatedly criticized that whole genre as a morass of ill-founded conjectures, and instead he advocated methodological caution and theoretical restraint in all the sciences.

These apparent paradoxes are clarified, and even perhaps resolved, when what is taken into account is the whole range of Cuvier's geological and paleontological research, of which a representative selection of texts is printed in this volume. Few historians or scientists have read Cuvier's work for themselves; many of those few have read only his "Preliminary discourse"; and of those, most of the anglophones have read only Jameson's edition. The "Discourse" (text 19) is not inconsistent with the rest of Cuvier's work; but like the popular lectures (chapter 8) from which it was derived, it was atypical in one crucial respect: it was designed for a public much wider than that of the savants to whom the bulk of his research was directed. In the "Discourse" Cuvier allowed himself the luxury of a few flights of purple prose; but even here his speculation was restrained within what he believed could be demonstrated on clear evidential grounds.

The paradoxes just summarized can be evaluated briefly in turn. To begin with the supposed influence of Cuvier's religious beliefs: until recently few commentators stopped to consider just what concrete historical evidence exists about those beliefs.[2] That Jameson and other anglophone contemporaries made him out to be a staunch defender of Genesis is unquestionable; but they had their own reasons for wanting to recruit such a heavyweight savant onto their side in their own controversies, and they did so without active encouragement from the great man himself. Cuvier did indeed act in an official capacity in relation to the Protestant churches in France; but he did so late in life, at the height of his political

2. The outstanding and honorable exception is Outram, *Georges Cuvier* (1984): see p. 142ff.

power and influence, when he knew he could help defend the political and social rights of the small minority in which his cultural roots lay.[3]

It is notoriously difficult to assess the strictly *religious* commitments of any historical figure, and particularly of one who was as private as Cuvier about personal matters of any kind. Such evidence as there is, however, suggests that he was in this respect a typical example of a late Enlightenment savant, and that his commitments were decidedly formal. Although nominally a Protestant, he married a Roman Catholic. His reputation around that time—which was also the time of his most creative work on fossils—was as a religious skeptic, or even an atheist (text 10). His handling of internecine disputes within the Protestant churches, when much later that became one of his official duties, was studiously impartial, and he showed little patience with the finer points of doctrinal argument. Finally, and also late in his life, his surviving daughter—who was certainly a devout Protestant and must have known him as well as anyone—is recorded as having been in the habit of praying for her father's conversion.

None of this suggests a man of evangelical zeal. On the other hand, such religious formalism was quite compatible—as it was in many another Enlightenment savant—with an appreciation of the value of religious institutions as a cohesive and stabilizing force within society. In that sense Cuvier may well have regarded himself as a defender of religion. But he was clearly no literalist, nor—to use a grossly anachronistic term—a fundamentalist. In his popular lectures he apparently noted in passing the structural correspondence between the Creation narrative in Genesis and the geological record of the progress of life; but it is not clear how far he advocated that parallel as his own view, and probably he only mentioned it in order to make the new geology more palatable to the religious elements in his audience and to defend it against the religious conservatism that was in the ascendant at the time.

Like all other savants working in or around the field of "geology" at this time, Cuvier took it for granted that the timescale of earth history was vast beyond human comprehension, so that the Genesis story would have to be interpreted—if at all—in a highly figurative manner. But this too was no problem for Cuvier, since he was evidently well aware of the work of contemporary German scholars, whose new biblical criticism had shown the value—not least, the *religious* value—of analyzing the

3. Strictly speaking, Cuvier belonged to a minority within a minority: although himself a Lutheran, he was responsible for the civic affairs of the much larger number of Reformed (Calvinist) congregations in France. All the Protestants together, however, amounted to only about 2% of the total population.

Bible with the same historicist and contextualist methods as any other ancient text.

Cuvier's lengthy discussion of *all* the ancient texts that bore on the question of the historicity of the most recent "catastrophe" showed the same impartiality (text 19, sec. 36). All—including the story of Noah's Flood—were in his opinion highly corrupted accounts that could not be taken as literal historical fact. Conversely, however, he argued that they were all worth examining, to find the core of historicity that remained when the fabulous or legendary layers were peeled away. What then remained, he claimed, was a body of convergent textual evidence that the earth's surface had indeed been ravaged by a "catastrophe" of some kind, only a few thousand years ago, back in the infancy of human civilization if not of humanity itself. Cuvier clearly regarded that textual evidence as valuable support for his claim that the *natural* evidence of fossil bones also pointed to some such event. It is not clear, however, that he ever regarded it as having more than that complementary or even merely supplementary role. The multicultural textual evidence simply reinforced and confirmed the natural evidence; it established the last catastrophe as a decisive event on the borderline, as it were, between the "present world" and the "ancient world," between the human world and the prehuman, or, more precisely, between the civilized literate world and all that had preceded it. At this point, the natural sciences could collaborate with the human sciences, and the somewhat artificial unity of the Institut National in Paris could become a reality in the work of one of its most prominent members.

The origin of the human species was left entirely vague in Cuvier's writings. The last catastrophe had destroyed whatever forms human life and human societies had taken previously; or at least it had made the evidence inaccessible, perhaps beneath the present seas. In any case, there was no authentic *evidence* of human fossils, and for Cuvier that was what mattered. Without evidence speculation was fruitless, and Cuvier would not indulge in it, at least not in his public role as a savant.

Likewise Cuvier was extremely cautious in his wording about the origins of every other animal species, and indeed of larger groups such as the reptiles or the mammals. Until his time the question of the origin of species had hardly existed as a *scientific* problem. To ask about the origin of, say, the elephant species had been like asking about the origin of the "species" of, say, iron or common salt: significantly, the same word was used in both contexts. In both cases it was a question of natural entities that were simply part of the diversity of the natural world. Such entities could be described and classified, and that was indeed the primary task of "natural history"; but the *origin* of those entities was a matter of meta-physics or perhaps of theology, but certainly not of natural science. Only around Cuvier's time, as those who were later to be known as geologists began to produce clear evidence for a true *history* of the earth and of life, did the question of the origin of species or larger groups of organisms become a strictly scientific issue; and then only because and insofar as it became clear that such species or groups had not always existed, and therefore must have had a point of origin *in time*.

For Cuvier, in fact, the issue was even closer to home. Although the fossil shells found in older formations of strata were known to be very different from the shells of living mollusks, he was aware that the fossil species might well be living undiscovered in some remote part of the ocean. Only for terrestrial vertebrates, he argued, could one be more confident that fossils provided a reliable record of the history of life. So he maintained that evidence was only just emerging—in his own time and of course not least through his own work—to suggest that there had been a definite point in time at which, for example, reptiles "began to ex-ist," and another later point at which mammals had "appeared" (text 19, sec. 32). Of course this new fossil evidence put the question of origins squarely into the scientific realm; but Cuvier studiously abstained from any causal speculations about it, and confined himself to strictly phe-nomenal language, as in the phrases just quoted.

Doubtless one reason for Cuvier's reticence about origins was his hos-tility to the materialistic implications of the only kind of causal explana-tion under discussion at the time, namely Lamarck's transformist theo-ries. It is of course possible that Cuvier privately believed in some kind of supernatural causation for new species or for the origins of reptiles and mammals, but there is no historical evidence for it. It is more likely that he simply believed that Lamarck's was not the *correct* natural explanation, or even the correct *kind* of natural explanation; and that beyond that he declined to speculate, because without adequate evidence theorizing was valueless.

Cuvier was much more concerned with the extinctions of species than with their origins; and of course it was here that he became embroiled, perhaps unexpectedly, in matters of "geology." For he seems never to have considered the possibility that a species might become extinct grad-ually, by slowly losing a long battle to maintain its numbers in a natural habitat. Any species was for him such a well-adapted "animal machine" that only a *catastrophic* event could make it go extinct: under the impact of any more gradual changes in the environment, a well-adapted species would simply migrate and survive. In fact for Cuvier, as the texts in this volume show repeatedly, species did not just become "extinct" *[éteints]*—though he did sometimes use that word—but, rather, were "destroyed" *[détruits]* or "wiped out" *[anéantis]*.

Cuvier's inclination toward—even, commitment to—explanations in terms of catastrophic events is clear from his earliest geological writings, at least from the time he arrived in Paris (text 3). He already knew of Deluc's catastrophist theorizing (text 2), and probably derived his ideas in the first instance from that source; but that explains little, for he also knew of many other theories in which any notion of catastrophic events in the earth's past history was firmly rejected. All were equally available to him as potential *resources;* that he found catastrophist explanations the most persuasive, and chose to develop them on the basis of his own specific research, requires further explanation. Anyway he was correct in claiming (text 19, sec. 8) that his emphasis on catastrophes was relatively novel, at least in Paris, even allowing for Deluc's and Dolomieu's earlier work; certainly it was not a rearguard defense of a reactionary position.

Cuvier is unlikely to have been attracted to catastrophist explanations in order to bolster the historicity of Genesis for religious purposes. As was already suggested, the converse is more probably the case, namely that he deployed the evidence of Genesis—along with many other ancient texts—in order to bolster his *scientific* case for the reality and recent date of the physical event that, he claimed, had destroyed the entire fauna of the "ancient world." Deluc did indeed argue along somewhat similar lines for explicitly religious purposes; but it was widely recognized at the time that the merits of that scientific case were not dependent on Deluc's apologetics.

Another explanation of Cuvier's clear inclination toward catastrophes is quite plausible, but can hardly be assigned more than a supplementary or supportive role. The verbal resonance between Cuvier's "revolutions of the globe" and the political Revolution that he lived through as a young man is less striking than it may seem at first sight, since the word "revolution" was still generally used in a wide sense that in no way implied sudden or violent events. In the sciences it was used, for example, to describe the slow, regular and predictable movements of the planets; and Lamarck used it in his geology to describe what he maintained were equally slow and tranquil changes in the distribution of continents and oceans. A "revolution," in the physical world or in the human, was a major change, but not necessarily a sudden, let alone a violent, one.

However, Cuvier did attribute suddenness and violence to some (not all) of the "revolutions of the globe," namely to those he termed "catastrophes," and above all to the most recent of such events. It is indeed possible that the idea of a geologically recent catastrophe in the natural world became much more plausible to Cuvier in the light of his own apparently traumatic experience of the social catastrophe of the Terror. Although what he witnessed at first hand in the provinces was minor compared to what took place in Paris, the effects of the Revolution were still apparent all around him when he reached the capital, not least in the disruption of scientific activity, from which its institutions were only just beginning to recover.

Some such link between the social world and the natural, between catastrophic "revolutions" in earth history and the catastrophe of *the* Revolution in France, is perhaps suggested indirectly by Cuvier's own writing. Just as he conceived the natural catastrophe of recent earth history as an event that had not merely made many fossil mammals extinct, but had "destroyed" them, so he often chose the same rather striking verb *[détruire]* to express *social* processes, at least within the sciences: for example, the deployment of reliable factual evidence would serve to "destroy" ill-founded speculative "systems" in geology. So in Cuvier's mind—as in the minds of other scientists at other periods—there may have been a tacit isomorphism between the structure and dynamics of the social world and those of the natural. But this can hardly count as an adequate explanation of Cuvier's catastrophist geology, if only because other French savants did not replicate the political Revolution in their scientific work: Lamarck, for example, was more directly affected by the Terror than Cuvier, yet he articulated a vision of earth history utterly removed from any catastrophism.

A more adequate understanding of Cuvier's catastrophism must therefore be sought within his scientific project itself. For it is clear that his notion of a "catastrophe" was of a physical event that had had a natural cause of some kind. Catastrophes were part of the "order of nature." That was shown most clearly by the fact—as Cuvier claimed it to be—that catastrophes had occurred repeatedly in the course of earth history. Cuvier here parted company from Deluc, who had shown little interest in any but the most recent of such events. Far more significantly, Cuvier based his case on the detailed empirical work he had done with Brongniart around Paris, and particularly on his collaborator's recognition of the alternation between freshwater and marine sediments there. For that alternation showed that major changes in physical geography—as the two authors believed them to be—had affected the continents several times in succession.

Even with such changes established as part of the course of nature, however, it was not obvious from the field evidence that they must have been sudden events; indeed, that was the point on which the Parisian geology of Cuvier and Brongniart was cogently criticized in later years. That Cuvier attributed a catastrophic character to these changes therefore still requires some explanation. The simplest is that such an attribution put them into the same class as the most recent catastrophe, and therefore confirmed that such events were indeed part of the course of nature. But that brings us back to the problem of the "last revolution": it is clear that this was always central in Cuvier's conception of earth history, but it is not immediately clear why he claimed it had been a "catastrophe."

Cuvier's primary evidence for that claim was, of course, his fossil bones. Beginning as a digression—unexpected and unplanned—from his program of work on the comparative anatomy of living animals, this research became his main preoccupation, to the extent that he shelved his other magnum opus until *Ossemens fossiles* was safely completed (text 18). In earlier years, his published lectures on comparative anatomy were highly regarded throughout the scientific world; but Cuvier built his career and his reputation primarily as—in his own words—"a new species of antiquarian" who had recovered a whole fauna of extinct mammals from their fragmentary fossil bones (text 19, sec. 1).

The striking impact of Cuvier's work depended crucially on his claim that these animals really were extinct. That in turn depended on his claim that his kind of rigorous osteological research could distinguish reliably between species as similar as the mammoth and the living elephants. Once that technical expertise was conceded, his claim about extinction was only threatened by those who denied the reality of *any* specific distinctions, and who argued that any one species could have been transformed in time into another. Therefore the plausibility of the conclusions on which Cuvier's career was built depended in part on an adamant rejection of Lamarck's (or any other) transformism. But that rejection entailed—and was perhaps motivated by—Cuvier's equally clear alternative concept of each species as a stable and functionally integrated "animal machine" that was well adapted to a specific mode of life (texts 5; 19, sec. 30). Such a species could never, in his view, become extinct merely by slow changes in the habitats available. Hence the plausibility of his claim to have restored a complete fossil fauna depended not only on rejecting transformism but also on an equally adamant assertion that these distinct and distinctly adapted species had been *destroyed* by a sudden catastrophe.

This intimate link between Cuvier's work on fossil anatomy and his interpretation of the "last revolution" alone seems adequate to account for his tenacious advocacy of catastrophist explanations in geology. Only the assertion of a drastic and above all a *sudden* physical change could, in his opinion, guarantee his claim to have recovered a whole fauna of *extinct* mammals distinct from living species. And only that in turn could establish his still grander claim to have "burst the limits of time" by making that vanished "ancient world" reliably knowable to human beings, as a wholly "other" period in a true *history* of life (text 19, sec. 1).

Compared to that central claim, the identification of the physical *cause* of the catastrophic event was of secondary importance. Cuvier was in fact quite vague and inconsistent, even in his speculations about its physical character: as a transient marine incursion, as a sudden refrigeration, or as a major interchange between continents and oceans. Still less was he prepared to indulge in speculations about its physical cause, though he clearly regarded it as a wholly natural event.

For all the florid prose of some parts of his "Preliminary discourse," Cuvier was consistent in his restraint about causal explanations in geology. He argued that the presently observable processes known to him—admittedly an impoverished selection—were quite inadequate to account for the observable effects of the "last revolution." He therefore inferred that that event must have had a cause of some other kind, but he refused to speculate on what it might have been. There is no good reason to infer that this was designed to leave the door open for some kind of supernatural causation: that would have been to go against the spirit of his whole scientific endeavor. Conversely, Cuvier's own writings give every reason to conclude that his refraining from causal speculation in geology was rooted in his conception of what made for good and fruitful scientific work.

As the texts in this volume show, Cuvier was at first highly critical of those who were just beginning to call themselves "geologists," precisely because their causal speculations were too loosely tied to empirical evidence to yield fruitful results. The proliferation of the "theories of the earth" proposed by geologists, the sheer diversity of their "systems," were to Cuvier sure signs of too many theories chasing too few facts. But his was no naive empiricism; observations became what he termed "facts" only when they were reliably interpreted. Nonetheless, he maintained that far more empirical observation was needed, in order to constrain the conditions of plausibility of geologists' "systems" and to identify those that had some chance of being at least partly correct. Until such empirical research was done, the proliferating "systems" simply brought geology into disrepute, and undermined its claims to be considered a true science. Cuvier used every opportunity to promote his comparative anatomy as a science as rigorous as the physical and mathematical sciences; so he was not going to jeopardize his hard-won status by involving himself in geology, unless it too could be practiced in a truly scientific manner.

That possibility became apparent to Cuvier, and he stopped his criticism of "geology" and "geologists," only when he became aware of the distinctly different kind of work practiced in Germany under the name of *Geognosie* (text 14), and set himself with Brongniart to develop that model further in the Paris region (text 15). He then came to appreciate that there was, as it were, a halfway house between the atheoretical compilation of observations and the construction of an all-embracing global "system" for the "theory of the earth." That recognition led directly to his formulation of what was explicitly an *agenda* for geology, which would lead out of the impasse of a superfluity of overambitious theories, into a more modest but also more fruitful program for future research (texts 10; 19, secs. 22–25, 40).

In that perspective, Cuvier could safely leave to the future even such major questions as the causation of the "last revolution." Recognizing the special character of a *historical* natural science, such as geology was becoming, Cuvier realized the difference between establishing the historical reality and physical character of a past event, and determining its natural cause. That he limited himself to the former is no sign of his defective conception of science, still less that he was a covert supernaturalist; on the contrary, it is a sign of his highly sophisticated understanding of what was then a novel kind of science.

In conclusion, Cuvier's enduring legacy to geological science lay not so much in his catastrophism, important though that was in the nineteenth century, but rather in his rigorous and painstaking analysis of fossils. Here his method was actualistic through and through: his determination of the character of the extinct vertebrates depended wholly on his comparisons between their fossil bones and the skeletons of extant species preserved in the great museum in Paris where he made his career and his home. Having identified the fossils as truly distinct species, and inferentially as extinct ones as well, Cuvier at least foresaw the possibility of reconstructing them as they would have been when alive in their appropriate habitats. Although he himself never went far down that path, his ambition to do so was explicit from an early stage, and he did go some way when the opportunity presented itself (texts 5, 7). Such reconstructions of the animals of the "ancient world" were the most vivid expression of his ambition to demonstrate that reliable human knowledge of the prehuman world was not unattainable. The best guarantee of such knowledge was his demonstration of the sheer "otherness" of the animal world he had discovered; it was not a mere variant of the present but a truly different "ancient world." A real *history* of life on earth was within human grasp: the "new species of antiquarian" could indeed "burst the limits of time."

FURTHER READING

Full details of the publications mentioned below are given in the "Bibliography of Works by Historians of Science."

Three books, all in English, together provide an excellent basis for further study of Cuvier. The biographical narrative that links the texts printed in this volume has drawn on them so extensively that it would have been pointless to cite them repeatedly in the notes.

Coleman's *Georges Cuvier zoologist* (1964) was the first major study to make full use of the rich collections of Cuvier manuscripts preserved in Paris. It was also the first modern work that delved behind Cuvier's mythic reputation as the big baddie of nineteenth-century opposition to organic evolution, and tried to understand his work on its own terms rather than as ammunition for modern biological controversies. As the title implies, however, it focuses on Cuvier's zoological work, and specifically on the issue of transformism (as evolutionary theories were then known) and his conception of the animal organism and the nature of animal classification. One chapter deals with Cuvier's paleontological research, but that work is not adequately embedded in its contemporary geological context. Nonetheless, this is a fine starting point.

Negrin's *Georges Cuvier: Administrator and educator* (1977), with a title clearly echoing Coleman's, set out to complement the earlier work with a study of Cuvier's *non*scientific career; it too was based on extensive archival research. It presents a very clear description of his background and personal life, and a fine analysis of his many official positions and duties. Although the work disclaimed any ambition to contribute to a knowledge of Cuvier's scientific career, it was in this respect much too modest: the book is quite indispensable for understanding the immediate context in which he carried out his research. It is also invaluable for its perceptive and balanced assessments of the more personal aspects of Cuvier's life. It is far less well known than it deserves, because it was a doctoral dissertation that was never formally published in book form; it is however easily available.

Outram's *Georges Cuvier: Vocation, science and authority in post-Revolutionary France* (1984), finally, comes as near as any work to a fully rounded study of Cuvier's life and work as a single integrated whole. Like any other modern historical study of any value, its intensive use of manuscript sources goes without saying. Although it does not claim to be a biography, it follows Cuvier's career in broadly chronological phases. The book delves behind the myth—sedulously fostered by Cuvier himself and by his earlier biographers—of an effortless rise to distinction. The focus is on Cuvier's often laborious *construction* of a scientific career, at a period when such careers—as understood in the modern world—did not yet exist, and when networks of patronage and personal alliances counted for far more than formal appointments or duties. However, as with much modern historical analysis in this mode, the scientific research itself is not analyzed in any great detail.

A work of a quite different kind is indispensable for understanding Cuvier's published work and its dissemination. Smith's *Georges Cuvier: An annotated bibliography* (1993) lists Cuvier's published papers, including their translations and summaries in other languages, and thereby makes it possible to trace how, when, and where his work could be read by those without access to the originals. Complementing that work of reference is Outram's *Letters of Georges Cuvier* (1980), listing most of the surviving letters *from* Cuvier to others. Conversely, Dehérain's *Manuscrits du fonds Cuvier* (1908–22) is a chronological list, with summary contents, of the richest collection of letters *to* Cuvier. Combined, these works give a good impression of the range of his network of correspondents throughout Europe and even beyond.

Several other historical studies are valuable for what they describe about Cuvier and his science, although their primary focus is on his main antagonist Lamarck, and particularly on Lamarck's transformist theories: Burkhardt's *Spirit of system* (1977) and Corsi's *Age of Lamarck* (1988) are outstanding examples. Those who read French have a wider choice: Balan's *L'ordre et le temps* (1979) and Laurent's *Paléontologie et évolution*

(1987), though widely different in approach, are both important here; and Daudin's classic *Cuvier et Lamarck* (1926) remains invaluable.

In all this historical literature, the geological dimension of Cuvier's work remains under-researched and therefore underestimated; that is of course a major reason for the present work. Much excellent research that is relevant to the earlier, eighteenth-century context of Cuvier's geology is still scattered in articles in scholarly periodicals; the best surveys of the issues, again in French, are Gohau's *Sciences de la terre* (1990) and the second volume of Ellenberger's *Histoire de la géologie* (1994). The slightly later period in which Cuvier was most active remains underexplored.

An anthology such as this can whet the appetite; only the original works can or should satisfy it fully. Copies of some of the original printed sources can be found in any major research library, because they are not—relatively speaking—rare items. For those not lucky enough to have access to such a collection the choice is more limited; however, several of the relevant items have been republished in modern times, some in facsimile form. Even for those who read little or no French, handling the fine facsimile edition of the *Ossemens fossiles* (1969), for example, and seeing more of the illustrations than can be reproduced here, is an experience that adds immeasurably to an understanding of Cuvier's science. Since so many of Cuvier's special papers (see texts 7, 8, 11, 12, 15, and 16) were reissued in this work, there is material here for delving more deeply into many aspects of his work on fossils. The same work contains the famous "Preliminary Discourse" (text 19) in its original format, and there is an inexpensive modern edition (1992) of the same text. The first and third of Jameson's editions of the "Discourse" in translation are available in facsimile (1971 and 1978 respectively), and are valuable for showing how the reception of Cuvier's geology in the English-speaking world was subtly—or not so subtly—distorted by Jameson's prefaces and notes. Finally, Cuvier's *Rapport* on the progress of the sciences is also available in a facsimile reprint (1968); even with a minimal knowledge of French, the contemporary cognitive "map" of all the sciences—in other words, the context of Cuvier's review of geology (text 14)—can readily be appreciated. References to all these *modern* editions and reprints are given at the end of the relevant entries in the "Bibliography of Cuvier's Sources."

This list gives the full reference for each publication in its original language (works by classical Greek authors are given in Latin, as they are by Cuvier). Numbers in brackets after most entries indicate the *text(s)* in which the publication is cited by Cuvier (those in the form "19.20" denote, for example, section 20 of the long text 19); a question mark indicates an allusion rather than an explicit citation, or some other reason for uncertainty.

Also included are those of Cuvier's publications that are translated (in whole or in part), or from which illustrations are reproduced, in this volume; the relevant text is shown by a number in bold type after the abbreviated title. Also listed, for convenience, are primary sources mentioned in the editorial introductions and footnotes, but not specifically cited by Cuvier.

Adelung, Johann Christoph. 1806–17. *Mithridates, oder allgemeine Sprachenkunde mit dem Vater Unser als Sprachprobe in bey nahe fünfhundert Sprache und Mundarten.* 4 vols. Berlin. [19.36]

Aelian. *De natura animalium.* [19.28]

Agatharcides. *Excerptae historiae.* [19.28]

André, Noël. 1806. *Théorie de la surface actuelle de la terre.* Paris.

Aristotle. *Historia animalium.* [19.28]

Athenaeus. *Deipnosophistae.* [19.28]

Bentley, J. 1799. On the antiquity of the Suryá Siddhánta, and the formation of the astronomical cycles therein contained. *Asiatick researches* 6:537–88. [19.36]

————. 1805. On the Hindu systems of astronomy, and their connection with history in ancient and modern times. *Asiatick researches* 8:193–244 and table. [19.36]

Bertrand, Louis. 1799. *Renouvellements périodiques des continens terrestres.* Paris. [19.20]

Breislak, Scipio. 1801. *Voyages dans la Campanie.* 2 vols. Paris. [14]

Brémontier, Nicolas Théodore. 1797. Sur les moyens de fixer les dunes qui se trouvent entre Bayonne et la pointe de Grave. *Journal de l'École Polytechnique* 2:61–70. [19.35]

Brocchi, Giovanni Battista. 1814. *Conchiologia fossile subappenina con osservazione geologiche sugli Appenini e sul suolo adiacente.* 2 vols. Milan.

Brongniart, Alexandre. 1807. *Traité élémentaire de minéralogie, avec des applications aux arts; ouvrage destiné à l'enseignement dans les lycées nationaux.* 2 vols. Paris. [15]

————. 1810. Sur les terrains qui paraissent avoir été formés sous l'eau douce. *Annales du Muséum d'Histoire Naturelle* 15:357–405. [15]

Buffon, Georges-Louis de. 1778. Des époques de la nature. *Histoire naturelle* 20 (supplément 5):1–254. [3, 13?, 19.19]

Burnet, Thomas. 1680–89. *Telluris theoria sacra: Orbis nostri originem & mutationes generales, quae aut jam subiit, aut olim subiturus est, complectens.* 2 vols. London. [19.19]

Calpurnius. *Eclogae.* [19.28]

Capitolinus, Julius. *Gordiani tres.* [19.28]

Chateaubriand, François-Auguste-René. 1802. *Génie du Christianisme, ou beautés de la réligion chrétienne.* Paris.

Colebrooke, H. T. 1805. On the Vedas, or sacred writings of the Hindus. *Asiatick researches* 8:369–476. [19.36]

Coupé, Jacques Michel. 1805. Sur l'étude du sol des environs de Paris. *Journal de physique* 61:363–95. [15]

Ctesias. *Indica.* [19.28]

Cuperus, Gisbertus. 1719. *De elephantis in nummis obviis exercitationes duae.* The Hague. [19.28]

Cuvier, Frédéric. 1811. Recherches sur les caractères ostéologiques qui distinguent les principales races du chien domestique. *Annales du Muséum d'Histoire Naturelle* 18:333–53, pls. 18–20. [19.33]

Cuvier, Georges. 1796a. Mémoire sur les éspèces d'elephans tant vivantes que fossiles, lu à la séance publique de l'Institut National le 15 germinal, an IV. *Magasin encyclopédique,* 2e année, 3:440–45. [3]

————. 1796b. Notice sur le squelette d'une très-grande espèce de quadrupède inconnue jusqu'à présent, trouvé au Paraguay, et déposé au cabinet d'histoire naturelle de Madrid, redigée par G. Cuvier. *Magasin encyclopédique,* 2e année, 1:303–10, 2 pls. [4]

————. 1798. *Tableau élémentaire de l'histoire naturelle des animaux.* Paris.

————. 1799. Mémoire sur les espèces d'éléphans vivantes et fossiles, lu le premier pluvose an 4 [21 January 1796]. *Mémoires de l'Institut National des Sciences et des Arts, sciences mathématiques et physiques* (mémoires) 2:1–22, pls. 2–6. [3]

————. Year IX [1800]. Extrait d'un ouvrage sur les espèces de quadrupèdes dont on a trouvé les ossemens dans l'intérieur de la terre, addressé aux savants et aux amateurs des sciences: Imprimé par ordre de la classe des sciences mathématiques et physiques de l'Institut National, du 26 brumaire an 9 [16 nov 1800]. *Journal de physique, de l'histoire naturelle, et des arts* 52:253–67. [6]

————. 1800–1805. *Leçons d'anatomie comparée.* 5 vols. Paris.

————. 1804a. Mémoire sur le squelette presque entier d'un petit quadrupède du genre de sariges, trouvé dans le pierre à plâtre des environs de Paris. *Annales du Muséum d'Histoire Naturelle* 5:277–92, pl. 19. [8]

————. 1804b. Mémoire sur l'ibis des anciens Egyptiens. *Annales du Muséum d'Histoire Naturelle* 4:116–35, pls. 52–54. [19]

————. 1804–8. Sur les espèces d'animaux dont proviennent les os fossiles répandus dans la pierre à plâtre des environs de Paris. *Annales du Muséum d'Histoire Naturelle* 3:275–303, 364–87, 442–72; 4:66–75; 6:253–83; 9:10–44, 89–102, 205–15, 272–82; 12:271–84. [7]

————. 1806a. Sur les éléphans vivans et fossiles. *Annales du Muséum d'Histoire Naturelle* 8:1–58, 93–155, 249–69, pls. 38–45. [11]

————. 1806b. Sur le grande Mastodonte, animal très-voisin de l'éléphant, mais à mâchelières hérissées de gros tubercles, dont on trouve les os en divers endroits des deux continens, et surtout près des bords de l'Ohio, dans l'Amérique Septentrionale, improprement nommé Mammouth par les Anglais et par les habitans des États-Unis. *Annales du Muséum d'Histoire Naturelle* 8:270–312, pls. 49–56. [12]

————. 1806c. Sur différentes dents du genre des mastodontes, mais d'espèces moindres que celles de l'Ohio, trouvées en plusieurs lieux des deux continents. *Annales du Muséum d'Histoire Naturelle* 8:401–24, pls. 66–69. [12]

————. 1808. Sur le grand animal fossile des carrières de Maestricht. *Annales du Muséum d'Histoire Naturelle* 12:145–76, pls. 19–20. [16]

————. 1809. Sur les os fossiles de ruminans trouvés dans les terrains meubles. *Annales du Muséum d'Histoire Naturelle* 12:333–98, pls. 32–34. [16]

————. 1810. *Rapport historique sur les progrès des sciences naturelles depuis 1789, et sur leur état actuel, présenté à sa Majesté l'Empereur et Roi, en son Conseil d'État, le 6 fevrier 1808, par la Classe des Sciences physiques et mathématiques de l'Institut, conformément à l'arrêté du gouvernement du 13 ventôse an X [4 March 1802].* Paris. Facsimile reprint, Brussels: Culture et Civilisation, 1968. [14]

————. 1812a. *Recherches sur les ossemens fossiles de quadrupèdes, où l'on rétablit les caractères de plusieurs espèces d'animaux que les révolutions du globe paroissent avoir détruites.* 4 vols. Paris. Facsimile reprint, Brussels: Culture et Civilisation, 1969. Reprint of "Discours préliminaire" in *Recherches sur les ossements fossiles de quadrupèdes: Discours préliminaire*, ed. Pierre Pellegrin. Paris: Flammarion, 1992. [7, 11, 12, 17, 18, 19]

————. 1812b. Sur un nouveau rapprochement à établir entre les classes qui composent le règne animal. *Annales du Muséum d'Histoire Naturelle* 19:73–84.

————. 1813. *Essay on the theory of the earth, with mineralogical notes, and an account of Cuvier's geological discoveries by Professor Jameson.* Edinburgh. Facsimile reprint, Farnborough, England: Greg, 1971.

————. 1817a. *Essay on the theory of the earth.* 3rd ed. Edinburgh. Facsimile reprint, New York: Arno, 1978.

————. 1817b. *Le règne animal distribué d'après son organisation, pour servir de base à l'histoire naturelle des animaux et d'introduction à l'anatomie comparée.* 4 vols. Paris. Facsimile reprint, Brussels: Culture et Civilisation, 1969. [18, 19.1, 19.20, 19.30]

————. 1821–24. *Recherches sur les ossemens fossiles, où l'on rétablit les caractères de plusieurs espèces d'animaux dont les révolutions du globe ont détruites les espèces.* Nouvelle édition, entièrement refondue, et considérablement augmentée. 5 vols. in 7. Paris. [18]

————. 1826. *Discours sur les révolutions de la surface du globe, et sur les changemens qu'elles ont produit dans le règne animal.* Troisième édition française. Paris and Amsterdam. **[16]**

————. 1845. *Briefe an C. H. Pfaff aus den Jahren 1788 bis 1792, naturhistorischen, politischen, und literarischen Inhalts: Nebst einer biographischen Notiz über G. Cuvier.* [Ed. W. F. G. Behn]. Kiel. **[1, 2]**

Cuvier, Georges, and Alexandre Brongniart. 1808. Essai sur la géographie minéralogique des environs de Paris. *Annales du Muséum d'Histoire Naturelle* 11:293–326. **[15,** 18]

————. 1811. Essai sur la géographie minéralogique des environs de Paris (lu 11 avril 1808). *Mémoires de la Classe des Sciences mathématiques et physiques de l'Institut Impérial de France* 1810:1–278, 2 pls., 1 map. **[15,** 18, 19.40]

Cuvier, Georges, René-Just Haüy, and Claude Hugues Lelièvre. 1807. Rapport de l'Institut National (Classe des Sciences physiques et mathématiques), sur l'ouvrage de M. André, ayant pour titre: Théorie de la surface actuelle de la terre. *Journal des mines* 21:413–30. **[13]**

Dacier, Bon-Josephe. 1810. *Rapport historique sur les progrès de l'histoire et de la littérature ancienne depuis 1789, et sur leur état actuel, présenté à sa Majesté l'Empereur et Roi, en son Conseil d'état, le 20 fevrier 1808.* Paris.

Deluc, Jean André. 1778. *Lettres physiques et morales sur les montagnes et sur l'histoire de la terre et de l'homme: Addressées à la Reine de la Grande-Bretagne.* The Hague.

————. 1779. *Lettres physiques et morales sur l'histoire de la terre et de l'homme: Addressées à la Reine de la Grande Bretagne.* 5 vols. in 6. The Hague and Paris. [2, 14, 19.35]

————. 1790–93. Lettres à M. de la Métherie. *Observations sur la physique, sur l'histoire naturelle, et sur les arts* 36:144–54, 193–207, 276–90, 363–79, 450–69; 37:54–71, 120–38, 202–19, 290–308, 332–51, 441–59; 38:90–109, 174–91, 271–88, 378–94; 39:215–30, 332–48, 453–64; 40:101–16, 180–97, 275–

92, 352–69, 450–67; 41:32–50, 123–40, 221–39, 328–45, 414–31; 42:88–103, 218–37. [2, 14]

———. 1798. *Lettres sur l'histoire physique de la terre, addressées à M. le professeur Blumenbach, renfermant de nouvelles preuves géologiques et historiques de la mission divine de Moyse.* Paris. [14]

Desmarest, Anselme-Gaétan. 1811. Mémoire sur la Gyroconite. *Nouvelle bulletin de la Société Philomathique* 2, 4e année: 275–77.

Desmarest, Nicolas. 1794–1828. *Géographie physique.* 5 vols. In *Encyclopédie méthodique.* Paris. [14]

———. 1804. Seconde mémoire sur la constitution physique des couches de la colline de Montmartre et des autres collines correspondantes. *Mémoires de l'Institut National, Sciences mathématiques et physiques* (mémoires) 5:16–54. [15]

Dietrich, P. F. de. 1786–89. *Description des gîtes de minerai, des forges, et des salines des Pyrénées, suivie d'observations sur le fer mazé et sur les mines des Sards en Poitou.* 3 vols. Paris. [14]

Dion Cassius. *Historiae Romanae libri XXV.* [19.28]

Dolomieu, Déodat de. 1783. *Voyage aux Iles de Lipari fait en 1781, ou notices sur les Iles Aeoliennes, pour servir à l'histoire des volcans; suivi d'un mémoire sur une espèce de volcan d'air, & d'un autre sur la température du climat du Malthe, et sur la différence de la chaleur réelle & de la chaleur sensible.* Paris. [14]

———. 1788. *Mémoire sur les Iles Ponces, et catalogue raisonné des produits de l'Etna; pour servir à l'histoire des volcans: Suivis de la description de l'éruption de l'Etna, du mois de Juillet, 1787.* Paris. [14]

———. 1791–92. Mémoire sur les pierres composées et sur les roches. *Observations sur la physique, sur l'histoire naturelle, et sur les arts* 39:374–407; 40:41–62, 203–18, 372–403. [19.20]

———. 1793. Mémoire sur la constitution physique de l'Egypte. *Observations sur la physique, sur l'histoire naturelle, et sur les arts* 42:41–61, 108–26, 194–215. [19.35]

———. 1795. Observations sur la prétendue mine de charbon de terre dite de la *Desirée*, commune de Saint-Martin-la-Garenne, districte de Mantes. *Journal des mines* 2:45–58. [15]

Dureau de la Malle, A. J.-C.-R. 1807. *Géographie physique de la Mer Noire, de l'interieur de l'Afrique, et de la Méditerranée.* Paris. [19.35]

Eichhorn, Johann Gottfried. 1803. *Einleitung in das Alte Testament.* Dritte verbesserte und vermehrte Ausgabe. 4 vols. Leipzig. [19.36]

Eusebius. *Praeparatio evangelica.* [19.36]

Faujas de Saint-Fond, Barthélémy. 1778. *Recherches sur les volcans éteints du Vivarais et du Velay; avec un discours sur les volcans brûlans, des mémoires analytiques sur les schorls, la zéolite, le basalte, la pouzzolane, les laves, & les différentes substances qui s'y trouvent engagées, &c.* Grenoble and Paris. [14]

———. 1784. *Minéralogie des volcans, ou description de toutes les substances produites ou rejetées par les feux souterrains.* Paris. [14]

————. Year VII [1799]. *Histoire naturelle de la montagne de Saint-Pierre de Maestricht.* Paris.

Forfait, Pierre. 1800. Über die Lage, die Lagunen, Häfen, und die Seewesen von Venedig. *Monatliche Correspondenz zur Beförderung der Erd- und Himmelskunde* 1:1–20, 91–101. [19.35]

Fortia d'Urban, Agricole de. 1807. *Histoire de la Chine avant le déluge d'Ogigès.* 2 vols. Paris. [19.38]

Fortis, Alberto. 1802. *Mémoires sur l'histoire naturelle et principalement sur l'oryctographie de l'Italie.* 2 vols. Paris. [14]

Gatterer, Joseph Christoph. 1786. Commentatio prima [et altera] de theogonia Aegyptiorum ad Herodoti L. II. cap. 145. *Commentationes Societatis Regiae Scientiarum Gottingensis* 7, *Historicae et philologicae classis:* 1–57. [19.36]

[Gazola, Giovambattista]. 1796. *Ittiolitologia veronese del Museo Bozziano ora annesso a quello di Conte Giovambattista Gazola e di altri gabinetti di fossili Veronesi con la versione latina.* Verona. [14]

Le Gentil de la Galaisière, Guillaume Joseph. 1779–81. *Voyage dans les mers de l'Inde, fait par ordre du Roi, à l'occasion du passage de Venus, sur le disque du Soleil, le 6 juin 1761, et le 3 du même mois 1769.* 2 vols. Paris. [19.36]

Guignes, Joseph de. 1770. *Le Chou-king: Un des livres sacrés du Chinois.* Paris. [19.36]

————. 1809a. Observations sur les sares des Chaldéens, et sur le nombre incroyable d'années qu'on assigne aux règnes de leurs premiers rois. *Histoire de l'Académie des Inscriptions et des Belles-Lettres* 47, *Mémoires de littérature:* 345–77. [19.36]

————. 1809b. Mémoire concernant l'origine du zodiaque et du calendrier des orientaux, et celle de différentes constellations de leur ciel astronomique. *Histoire de l'Académie des Inscriptions et des Belles-Lettres* 47, *Mémoires de littérature:* 378–434. [19.37]

Haüy, René-Just. 1801. *Traité de minéralogie.* 5 vols. Paris. [14]

Heeren, Arnold Hermann Ludwig. 1810. *Handbuch der Geschichte der Staaten des Alterthums, mit besonderer Rücksicht auf ihre Verfassungen, ihren Handel, und ihre Colonien.* 2nd ed. Göttingen. [19.28]

Hernandez, Francisco. 1649. *Historiae animalium et mineralium Novae Hispaniae liber.* Rome. [19.28]

Herodotus. *Historia.* [19.28, 19.35]

Humboldt, Alexander von. 1810. *Vues des Cordillères et monumens des peuples indigènes de l'Amerique.* Vols. 15 and 16 of *Voyages aux régions equinoxiales du Nouveau Continent.* Paris. [19.36]

Hutton, James. 1795. *Theory of the earth, with proofs and illustrations; in four parts.* 2 vols. Edinburgh. [19.20?]

Jablonski, Paul Ernst. 1750–52. *Pantheon Aegyptiorum, sive de diis eorum commentarius, cum prolegomenis de religione et theologia Aegyptiorum.* 3 vols. Frankfurt. [19.36]

Jefferson, Thomas. 1799. A memoir on the discovery of certain bones of a quadruped of the clawed kind in the western parts of Virginia. *Transactions of the American Philosophical Society* 4:246–59. [6?]

[Jones, William]. 1805. Sur les dieux de la Grèce, de l'Italie, et de l'Inde: Dissertation composée en 1784, et revue depuis. In *Recherches asiatiques, ou mémoires de la société établie au Bengale pour faire des recherches sur l'histoire et les antiquités, les arts, les sciences, et la littérature de l'Asie,* ed. A. Labaume, 1:162–213, pls. 11–24. Paris. [19.36]

Josephus. *Antiquitates Judaicae.* [19.36]

Justinus. *Historiae Philippicae, et totius mundi originibus.* [19.36]

Labaume, A., ed. 1805. *Recherches asiatiques, ou mémoires de la société établie au Bengale pour faire des recherches sur l'histoire et les antiquités, les arts, les sciences, et la littérature de l'Asie; traduit de l'anglois par A. Labaume: Revus et augmentés de notes, pour la partie orientale, philologique et historique par M. Langlès . . . , et pour la partie des sciences exactes et naturelles par les MM. Cuvier, Delambre, Lamarck, et Olivier, Membres de l'Institut.* 2 vols. Paris. [19.36]

Lamanon, Robert de Paul de. 1782. Description de divers fossiles trouvés dans les carrières de Montmartre près Paris, & vues générales sur la formation des pierres gypseuses. *Observations sur la physique, sur l'histoire naturelle et sur les arts* 19:173–94, pls. 1–3. [19.20]

Lamarck, Jean-Baptiste de. 1802a. *Hydrogéologie, ou recherches sur l'influence qu'ont les eaux sur la surface du globe terrestre; sur les causes de l'existence de bassin des mers, de son déplacement, et de son transport successif sur les différens points de la surface de ce globe; enfin sur les changemens que les corps vivans exercent sur la nature et l'état de cette surface.* Paris. [19.20]

———. 1802b. *Recherches sur l'organisation des corps vivans, et particulièrement sur son origine, sur la cause de ses développemens et des progrès de sa composition, et sur celle qui, tendant continuellement à la détruire dans chaque individu, amène nécessairement sa mort; précédé du discours d'ouverture du cours de zoologie, donné au Muséum National d'Histoire Naturelle, l'an X de la République.* Paris. [10?]

———. 1802–9. Mémoires sur les fossiles des environs de Paris, comprenant la détermination des espèces qui appartiennent aux animaux marins sans vertèbres, et dont la plupart sont figurés dans la collection des vélins du Museum. *Annales du Muséum d'Histoire Naturelle* 1:299–312, 383–91, 474–79; 2:57–64, 163–69, 217–27, 315–21, 385–91; 3:163–70, 266–74, 343–52, 436–41; 4:46–55, 151–57, 212–22, 289–98, 429–36; 5:28–36, 91–98, 179–88, 237–45, 349–57; 6:117–26, 214–28, 337–45, 407–15; 7:53–62, 130–39, 231–44, 419–30; 8:77–79, 156–66, 347–55, 383–88, 461–69; 9:236–40, 399–401; 12:456–59; 14:374–75. [14, 15]

———. 1804. Sur une nouvelle espèce de Trigonie, et sur une nouvelle espèce d'huître découverte dans le voyage du capitaine Baudin. *Annales du Muséum d'Histoire Naturelle* 4:351–59.

———. 1805. Considérations sur quelques faits applicable à la théorie du globe, observés par M. Péron dans son voyage aux terres australes, et sur quelques questions géologiques qui naissent de la connoissance de ces faits. *Annales du Muséum d'Histoire Naturelle* 6:26–52.

————. 1809. *Philosophie zoologique, ou exposition des considérations relatives à l'histoire naturelle des animaux; à la diversité de leur organisation et des facultés qu'ils en obtiennent; aux causes physiques qui maintiennent en eux la vie et donnent lieu aux mouvemens qu'ils exécutent; enfin, à celles qui produisent, les unes le sentiment, et les autres l'intelligence de ceux qui en sont doués.* 2 vols. Paris. [19.20]

Lamétherie, Jean-Claude de. 1797. *Théorie de la terre.* Seconde édition, corrigée, et augmentée d'une minéralogie. 5 vols. Paris. [13?, 14, 19.20]

————. 1804. *Considérations sur les êtres organisés.* 2 vols. Paris. [10?]

————. 1813. *Leçons de géologie, données au Collège de France.* 2 vols. Paris. [19.20?]

Laplace, Pierre Simon de. Year VII [1798–99]. *Exposition du système du monde.* Seconde édition, revue et augmentée par l'auteur. Paris. [19.36]

————. 1799–1805. *Traité de méchanique céleste.* 5 vols. Paris. [8, 19.37]

Lebrun, Corneille. 1718. *Voyage par la Moscovie en Perse et aux Indes orientales.* 2 vols. Amsterdam. [19.28]

Leibniz, Gottfried Wilhelm. 1693. Protogaea. *Acta eruditorum* [Leipzig] 1693: 40–42. [19.19]

————. 1749. *Protogaea sive de prima facie telluris et antiquissimae historiae vestigiis en ipsis naturae monumentis dissertatio.* Göttingen. [19.19]

Lichtenstein, Anton August Heinrich. 1791. *Commentatio philologica de simiarum quotquot veteribus innotuerunt, formis, earumque nominibus, pro specimine methodi qua historia naturalis veterum ad systema naturae Linnaeanum exigenda.* Hamburg. [19.28]

Lucian. *De dea syria.* [19.36]

Macrobius. *Somnium Scipionis.* [19.36]

Maillet, Benoît de. 1748. *Telliamed, ou entretiens d'un philosophe indien avec un missionaire françois, sur la diminution de la mer, la formation de la terre, l'origine de l'homme &c. mis en ordre sur les mémoires de feu M. de Maillet, par J. A. G****. 2 vols. Amsterdam. [19.19, 19.20]

Margravius, Georgius. 1648. *Historiae rerum naturalium Brasiliae libri 8.* Amsterdam. [19.28]

Marschall von Bibenstein, Karl Wilhelm, and Ernst Franz Ludwig Marschall von Bibenstein. 1802. *Untersuchungen über den Ursprung und die Ausbildung der gegenwärtigen Anordnung des Weltbäudes.* Giessen and Darmstadt. [19.20]

Marsham, John. 1672. *Chronicus canon Aegyptiacus, Ebraicus, Graecus.* London. [19.36?]

Montlosier, François Dominique de Reynaud de. 1789. *Essai sur la théorie des volcans d'Auvergne.* Paris. [14]

Onesicritus: see Strabo.

Oppian. *Cynergetica.* [19.28]

Palassou. 1784. *Essai sur la minéralogie des Monts-Pyrénées.* Paris. [14]

Pallas, Peter Simon. 1771–76. *Reise durch verschiedene Provinzen des Russischen Reichs.* 4 vols. St. Petersburg. [14]

———. 1778. Observations sur la formation des montagnes et les changemens arrivés au globe, particulièrement à l'égard de l'Empire de Russe. *Acta Academiae Petropolitanae* 1777 (1): 21–64. [14, 19.7]

Parkinson, James. 1804–11. *Organic remains of a former world: An examination of the mineralized remains of the vegetables and animals of the antediluvian world; generally termed extraneous fossils.* 3 vols. London. [14?]

Paterculus, Gaius Velleius. *Historiae Romanae libri duo.* [19.36]

Paterson: see Wilford.

Patrin, Eugène Louis Melchior. [1802–4]. [Unspecified articles.] In *Nouveau dictionnaire d'histoire naturelle.* 24 vols. Paris. [19.20]

Philostorgius. *Historia ecclesiastica.* [19.28]

Photius. *Bibliotheca.* [19.28]

Picot de la Peyrouse, Philippe. 1786. *Traité sur les mines de fer et les forges du comté de Foix.* Toulouse. [14]

Pini, Ermenegildo. 1790–92. Sulle rivoluzione del globo terrestre provenienti dall'azione dell'acque: Memoria geologica. *Memorie de matematica e fisica della Società Italiana delle Scienze* 5:163–258; 6:389–500. [2]

Playfair, John. 1802. *Illustrations of the Huttonian theory of the earth.* Edinburgh. [19.20]

Pliny the Elder. *Naturalis historiae.* [19.28]

Prony, Gaspard Riche de. 1812. Extrait des recherches de M. de Prony, sur le système hydraulique de l'Italie. In Cuvier, *Recherches sur les ossemens fossiles,* vol. 1, Discours préliminaire, 117–20. [19.35]

Ramond de Carbonnières, Louis François Elisabeth. 1789. *Observations faites dans les Pyrénées, pour servir de suite à des observations sur les Alpes, insérées dans une traduction des lettres de W. Coxe, sur la Suisse.* 2 vols. Paris. [14]

———. 1801. *Voyage au Mont-Perdu.* Paris. [14]

Rennell, James. 1800. *The geographical system of Herodotus examined and explained by a comparison with those of other ancient authors, and with modern geography....* London. [19.35]

Reuss, Franz Ambros. 1801–6. *Lehrbuch der Geognosie.* Vol. 3 of *Lehrbuch der Mineralogie, nach dem Herrn D. L. G. Karsten mineralogischen Tabellen ausgeführt.* 8 vols. Leipzig. [14]

Rodig, Johann Christian. 1801. *Lebende Natur.* Leipzig. [19.20]

Sale, George, et al., eds. 1780–84. *The modern part of an universal history, from the earliest accounts to the present time.* Compiled from original authors. By the authors of the ancient part. 42 vols. London. [19.36]

Saussure, Horace-Bénédict de. 1779–96. *Voyages dans les Alpes, précédés d'un essai sur l'histoire naturelle des environs de Geneve.* 4 vols. Neuchâtel. [14, 19.7]

[Scheuchzer, Johann.] 1709. Editorial report on letter "De montium structura." *Histoire de l'Académie des Sciences* 1708:30–33. [19.19]

Scheuchzer, Johann Jacob. 1726. *Homo diluvii testis et theoskopos.* Zurich.

Schlotheim, Ernst von. 1804. *Beschreibung merkwürdiger Kraüter- Abdrücke und Pflanzen-Versteinerungen: Ein Beitrag zur Flora der Vorwelt.* Gotha. [14]

Smith, W. 1815. *A delineation of the strata of England and Wales with part of Scotland.* London.

Solinus. *Memorabilia mundi.* [19.28]

Strabo. *Geographia.* Includes summaries of lost works by Onesicritus. [19.28]

Syncellus, Georgius. *Chronographia.* [19.36]

Tassin. Year X [1801–2]. *Rapport sur les dunes du golfe de Gascogne.* Mont-de-Marsan. [19.35]

Werner, Abraham Gottlob. 1787. *Kurze Klassifikation und Beschreibung der verschiedenen Gebirgsarten.* Dresden. [1?]

―――. 1802. *Nouvelle théorie de la formation des filons: Application de cette théorie à l'exploitation des mines, particulièrement de celles de Freiberg.* Paris. [14]

Whiston, William. 1696. *A new theory of the earth, from its original, to the consummation of all things, wherein the creation of the world in six days, the universal deluge, and the general conflagration, as laid down in the Holy Scriptures, are shown to be perfectly agreeable to reason and philosophy.* London. [19.19]

Wilford, F. 1807. Of the kings of Magad'ha: Their chronology. *Asiatick researches* 9:82–116 and table. [19.36]

Woodward, John. 1695. *An essay toward a natural history of the earth: And terrestrial bodies, especially minerals: As also of the seas, rivers, and springs; with an account of the universal deluge, and of the effects that it had upon the earth.* London. [19.19]

BIBLIOGRAPHY OF WORKS

BY HISTORIANS OF SCIENCE

The following publications are listed in "Further Reading," mentioned in the editorial introductions and footnotes, or used as sources for Cuvier's texts. Since this book is not intended to be a definitive study of his geological and paleontological research, let alone of his scientific work as a whole, references to historical research articles have been kept to a minimum. Needless to say, in my editorial sections and annotations I have made use of a large number of publications that are not listed here; I plan to make those debts explicit elsewhere.

Ardouin, P. 1970. *Georges Cuvier: Promoteur de l'idée évolutionniste et créateur de la biologie moderne.* Paris: Expansion Scientifique Française.
Balan, Bernard. 1979. *L'ordre et le temps: L'anatomie comparée et l'histoire des vivants au XIXe siècle.* Paris: Vrin.
Bultingaire, L. 1932. Iconographie de Georges Cuvier. *Archives du Muséum d'Histoire Naturelle,* ser. 6, 9:1–12, pls. I–II.
Burkhardt, Richard W., Jr. 1977. *The spirit of system: Lamarck and evolutionary biology.* Cambridge, Mass.: Harvard University Press.
Coleman, William. 1964. *Georges Cuvier zoologist: A study in the history of evolution theory.* Cambridge, Mass.: Harvard University Press.
Corsi, Pietro. 1988. *The age of Lamarck: Evolutionary theories in France, 1790–1830.* Berkeley: University of California Press.

Daudin, H. 1926. *Cuvier et Lamarck: Les classes zoologiques et l'idée de série animale, 1790–1830.* Paris: Felix Alcan.

Dehérain, H. 1908–22. *Catalogue des manuscrits du fonds Cuvier conservés à la Bibliothèque de l'Institut de France.* 2 vols. Paris and Hendaye.

Ellenberger, François. 1994. *Histoire de la géologie.* Vol. 2, *La grande éclosion et ses prémices, 1660–1810.* Paris: Technique et Documentation.

Gohau, Gabriel. 1990. *Les sciences de la terre aux XVIIe et XVIIIe siècles: Naissance de la géologie.* Paris: Albin Michel.

Grandchamp, Philippe. 1995. Deux exposés des doctrines de Cuvier antérieurs au "Discours préliminaire": Les cours de géologie professés au Collège de France en 1805 et 1808. *Travaux du Comité Français pour l'Histoire de la Géologie,* ser. 3, 8:13–26.

Laurent, Goulven. 1987. *Paléontologie et évolution en France de 1800 à 1860: Une histoire des idées de Cuvier et Lamarck à Darwin.* Paris: Comité des Travaux Historiques.

Negrin, Howard. 1977. *Georges Cuvier: Administrator and educator.* Ann Arbor: University Microfilms.

Outram, Dorinda. 1980. *The letters of Georges Cuvier: A summary catalogue of manuscript and printed materials preserved in Europe, the United States of America, and Australasia.* BSHS Monographs 2. Chalfont St. Giles: British Society of the History of Science.

———. 1984. *Georges Cuvier: Vocation, science, and authority in post-revolutionary France.* Manchester: Manchester University Press.

Plan, Danielle. 1909. *Un génévois d'autrefois: Henri-Albert Gosse, 1753–1816, d'après des lettres et des documents inédits.* Paris and Geneva: Fischbacher.

Rudwick, Martin J. S. 1985. *The great Devonian controversy: The shaping of scientific knowledge among gentlemanly specialists.* Chicago: University of Chicago Press.

———. 1992. *Scenes from deep time: Early pictorial representations of the prehistoric world.* Chicago: University of Chicago Press.

Russell, E. S. 1916. *Form and function: A contribution to the history of morphology.* London: John Murray. Reprint, Chicago: University of Chicago Press, 1982.

Smith, Jean Chandler. 1993. *Georges Cuvier: An annotated bibliography of his published works.* Washington, D.C.: Smithsonian Institution Press.

———————

Text 5: Cuvier's Public Lecture at the Institut, 1798

This manuscript, in Cuvier's hand, is in a folder marked "Supplément général à tous les mémoires imprimés de la 2me. partie," in the box MS 628, which is part of the Cuvier papers preserved in the Bibliothèque Centrale, Muséum National d'Histoire Naturelle, Paris. Numbers in brackets refer to the pages of the manuscript (not numbered by Cuvier). Correct accents have been added where necessary, and (within brackets) a few extra marks of punctuation.

[1] Extrait d'un Mémoire sur un animal dont on trouve les ossements dans la pierre à plâtre des environs de Paris, & qui paraît ne plus exister vivant aujourd'hui.

lu à la Séance publique de l'Institut national, du 15 Vendémiaire an 7.

Il n'est plus personne qui ne sache que la terre que nous habitons présente de toutes parts des traces manifestes de grandes et violentes révolutions; mais l'histoire de ces bouleversements n'a pu encore être débrouillée malgré les efforts de ceux qui en ont recueilli & comparé les documens.

[2] Les ossements de quadrupèdes qui se trouvent dans l'intérieur des couches qui composent nos continents, sont un des résultats les plus singuliers de ces révolutions.

L'examen approfondi qu'on en a fait dans ces derniers tems a montré qu'ils viennent presque toujours d'animaux étrangers au climat dans le quel on les découvre ou même d'animaux entièrement inconnus aujourd'hui.

Il faudra donc désormais ajouter à l'histoire des animaux qui existent présentement dans chaque contrée, celle des animaux qui y ont vécu, ou qui y ont été transportés autrefois. Pour cet effet il faudra que les [3] physiciens fassent pour l'histoire de la nature, ce que les antiquaires font pour l'histoire des arts et des moeurs des peuples; il faudra que les uns aillent chercher dans les ruines du globe les restes des êtres animés qui vivaient à sa surface, comme les autres fouillent dans les ruines des cités, pour y déterrer les monuments du goût du génie & des coutumes des hommes qui les habitaient.

Ces antiquités de la nature, si on peut s'exprimer ainsi, fourniront à l'histoire physique du globe, des monuments aussi utiles & aussi certains, que les antiquités ordinaires en fournissent à l'histoire politique et morale des nations.

[4] Mais ce n'est qu'à l'aide d'une connoissance rigoureusement exacte de l'anatomie comparée, que l'on pourra procéder à ces recherches sans redouter d'erreur; ce n'est que lors qu'on connoitra bien les squelettes de toutes les espèces vivantes qu'on pourra déterminer avec certitude si les ossements que la terre recèle proviennent ou non de quelques unes de ces espèces.

Aussi tant que l'anatomie comparée a été dans l'enfance on n'a donné d'attention qu'à ceux de ces ossements qui frappaient par leur grandeur ou par leur forme extraordinaire, encore les regardoit-on tantôt comme des os de géants, tantôt [5] comme des os d'Éléphants ou d'autres espèces connues.

Mais à mesure que cette partie de l'anatomie s'est perfectionnée, on a mis plus de précision dans cet examen, et Daubenton, Camper, & Pallas, sont ceux qui ont les premiers donné quelque chose d'un peu exact sur ce sujet.

Aujourd'hui l'anatomie comparée est parvenu à un tel point de perfection que l'on peut souvent d'après l'inspection d'un seul os, determiner la classe, quelquefois même le genre de l'animal auquel il a appartenu, surtout si cet os fait partie de la tête ou des membres.

Cette assertion n'étonnera point si on se rappelle que tous ces os dans [6] l'état de vie sont rassemblés en une espèce de charpente; que la place que chacun d'eux occupait est facile à reconnaitre, & qu'on peut juger par le nombre & la position de leurs facettes articulaires du nombre & de la direction des os qui leur étaient attachés.

Or le nombre, la direction & la figure des os qui composent chaque partie du corps d'un animal, sont toujours dans un rapport nécessaire, avec toutes les autres parties, de manière qu'on peut conclure, jusqu'à un certain point, de l'une d'elles à l'ensemble, & réciproquement.

Par exemple: lorsque les dents d'un animal sont telles [7] qu'il faut qu'elles soient pour qu'il se nourrisse de chair, nous pouvons assurer sans autre examen que tout le système de ses organes de la digestion est disposé pour cette sorte d'alimens[;] & que toute sa charpente & ses organes du mouvement, & même ceux de sa sensibilité sont disposés de manière à le rendre habile à poursuivre & à saisir une proye; car ces rapports sont les conditions nécessaires de l'existence de cet animal; et si les choses n'etaient pas ainsi, il ne pourroit pas subsister.

J'ai choisi cet exemple comme le plus palpable & le plus propre à vous donner une idée de la méthode qu'on employe dans [8] les recherches dont je vais vous entretenir. Vous sentez aisément que ces sortes de rapports entre les parties ne sont pas tous aussi évidents, & qu'à mesure qu'on déscend aux fonctions moins importantes on est reduit à des conjectures plus délicates, & à des conclusions moins certaines; mais il est du moins toujours facile d'assigner à chacun de ces résultats le degré de probabilité qui lui appartient.

Parmi les ossements que j'ai examinés d'après ces principes, les plus intéressans & les moins connus sont ceux qu'on trouve dans la pierre à plâtre [9] des environs de Paris.

Cette position dans l'intérieur même des immenses couches de gypse qui environnent cette ville du côte du nord, est déjà une circonstance singulière.

La plus part des débris de quadrupèdes que l'on a trouvés jusqu'ici se trouvent dans les couches très meubles, telles que des amas de sable, ou de limon qui ont pu être déposés par des rivières, ou bien dans des cavernes où ces animaux ont pu se retirer lors des inondations. Ceux dont je parle, au contraire, sont incrustés dans l'intérieur même de la pierre, & devraient se trouver déjà épars dans le liquide où elle s'est formée & [10] où elle les a saisis & enveloppés[.] Leur consistance est très friable, & ce n'est qu'avec beaucoup de précaution qu'on peut les en retirer. Ils sont ordinairement d'une teinte roussâtre. Leur abondance est telle qu'il n'est pas de jour que les ouvriers qui travaillent dans les carrières de Montmartre, de Mesnilmontant, de Pantin, d'Argenteuil & des autres villages environnans, n'en trouvent quelques uns dans les blocs qu'ils mettent en pièces; les vertèbres, les côtes et les dents isolées sont les morceaux les plus communs; les grands os [11] sont plus rares; surtout les mâchoires entières; & les os minces comme les omoplates, par ce qu'ils sont plus facile à briser.

Différens curieux de cette ville en receuillent depuis long-tems dans leurs cabinets & c'est en parcourant un grand nombre de ces collections que je me suis procuré les matériaux de ce mémoire. Celle qui m'en a fourni le plus avait été rassemblée par feu Joubert & appartient aujourd'hui au Citoyen Drée; auquel je dois beaucoup de reconnoissance pour la manière amicale dont il me les a communiqués.

Ayant donc examiné, décrit, dessiné & comparé près de cent [12] de ces morceaux, les ayant rapprochés les uns des autres suivant les indications que me donnaient leurs facettes articulaires, je suis parvenu à rétablir presque entièrement le squelette de l'animal auquel ils ont appartenu.

Je reserve pour une de nos séances particulières le détail et les preuves de toutes mes opérations qui j'accompagnerai des pièces qui leur ont servi de base, & je vais seulement en présenter ici le resultat, en vous donnant une idée du squelette de cet animal, [13] tel que les morceaux que j'ai examinés montrent qu'il devoit être.

Ses dents machelières ont des couronne plattes, qui présentent des compartimens de substance osseuse & de substance émailleuse. C'est la structure qu'elles ont dans tous les animaux qui se nourrissent d'herbe, parce qu'il leur falloit des espèces de meules pour broyer & non des espèces de ciseaux pour couper comme aux carnassiers.

La forme particulière de ces dents est assez semblable à ce qu'on voit dans le Rhinoceros[,] c'est à dire que les supérieures sont carrées & que les inférieures sont en double croissant [14] mais les incisives sont tranchantes au nombre de six à chaque mâchoire, suivies d'une canine de chaque coté derrière la quelle est un très court espace vuide.

Sans cette existence des canines & des incisives aux deux mâchoires on seroit tenté de prendre notre animal pour un ruminant tant ses dents machelières ressemblent à celles du cerf par leur surface extérieure; mais leur couronne est toute différente.

Cette disposition de ses dents est en général celle qu'on observe dans les cochons, les Tapirs, les Hippopotames, les Rhinoceros & les autre herbivores à cuir [15] épais dont les pieds sont terminés par plusieurs sabots.

Ainsi pour l'inspection seule de ces dents nous pouvons dejà juger que notre animal appartient à cette même classe. Nous allons voir que tout le reste de son squelette confirme cette conjecture.

La forme générale de sa tête, les courbures & les contours de ses différentes parties, ont tant de rassemblance avec celle du Tapir, qu'on est d'abord tenté de les regarder comme provenant de cet animal de l'amerique Méridionale.

Les os de nez & du museau sont même formés de manière qu'il paraît avoir eu aussi une courte trompe comme le Tapir.

Les pieds de devant ont trois [16] doigts apparents; ceux de derrière deux; cela se voit non seulement par les facettes des os de poignet & du cou de pied; mais encore par des morceaux de pierre où ces pieds se sont trouvés conservés en entier.

Ce nombre de doigts est d'autant plus remarquable que les naturalistes ne l'ont encore observé dans aucun quadrupède. Il achève de compléter les combinaisons possibles dans la classe à la quelle notre animal appartient; car, l'Éléphant en a cinq devant & cinq derrière; l'hippopotame & le cochon quatre devant et quatre derrière[;] le Tapir quatre devant & trois derrière[;] & notre animal trois devant & deux derrière, ce qui le place [17] immédiatement devant les animaux ruminans, qui ont deux doigts devant & deux derrière, & avec les quels nous venons de voir qu'il a encore quelques rapports par d'autres parties.

La découverte de l'animal indépendamment de son importance pour la théorie de la terre, sert donc encore à remplir une lacune dans l'échelle des êtres.

Ce que je viens de vous dire concernant les parties les plus importantes de son squelette suffit pour montrer qu'il diffère essentiellement de tous ceux que les naturalistes & les voyageurs ont découverts jusqu'à présent sur la surface du globe & c'est une preuve de plus du grand fait dont j'ai déjà une fois entretenu [18] le public, que plusieurs espèces d'animaux ont été entièrement détruites par les révolutions que notre planète a essuyées; ainsi je vous épargnerai la description plus détaillée des autres os qu'on ne pourroit d'ailleurs saisir sans l'inspection même des pièces.

Il ne seroit pas impossible[,] les os étant bien connus, de déterminer les formes des muscles qui s'y attachaient; car ces formes dépendent nécessairement de celles des os & de leurs éminences[.] Les chairs une fois rétablies il seroit aisé de se les figurer couvertes de leur peau, & on [19] auroit ainsi l'image, non seulement du squelette qui existe encore mais de l'animal entier, tel qu'il existoit autrefois. On pourroit même avec un peu plus de hardiesse deviner une partie de ses habitudes; car les habitudes d'un animal quelconque dépendent de l'organisation & en connoissant celle cy on peut conclure celles là; après tout ces conjectures ne serait peut-être guère plus hasardées, que celles que les Géologistes vont se trouver obligés de faire pour expliquer dans leurs systèmes comment les os d'un animal inconnu se trouvent dans un pays qui l'est tant. Et comment en effet ne pas pardonner quelques écarts à l'imagination, échauffée par un si grand spectacle? Comment réprimer ce désir si naturel de se rendre compte des causes qui ont pu produire de si terribles effets; élever les montagnes;

transporter les mers, détruire des espèces entières; changer en un mot la face du globe, & la nature des êtres qui l'habitent.

Mais on n'aime aujourd'hui dans les Sciences que ce qui se voit ou se calcule; on se soucie peu de ce qui se devine[;] & je me contenterai d'avoir ajouté quelques faits à la masse déjà si imposante que [20] les observateurs en ont rassemblée en montrant,

1° Que les ossements fossiles qui se trouvent dans le gypse des environs de Paris proviennent d'un animal très différent par ses formes de tous ceux qui habitent aujourd'hui dans notre climat.

2° Que cet animal ne s'est trouvé vivant dans aucun pays connu jusqu'à ce jour.

3° Qu'il forme un genre particulier qui doit se placer à la fin de la famille des Pachidermes, à la suite du Rhinoceros & du Tapir; & immédiatement avant le chameau qui commence la classe des ruminants.

Text 9: Cuvier's Notes for His Geology Lectures, 1805

This manuscript, in Cuvier's hand, is MS 3111 in the Bibliothèque de l'Institut de France, Paris (it was MS 111 in the catalog by Dehérain, *Manuscrits du fonds Cuvier*, 1908–22). The page numbering (in brackets) is that of the manuscript as now bound with others. Correct accents have been added where necessary, and (in brackets) a few marks of punctuation.

[56r] Cours du Lycée de l'an XIII Géologie

[57r] Plan général
C'est dans les couches à fossiles qu'est la plus forte preuve que le globe n'a pas toujours été comme à présent.

État des fossiles
simplement fossiles
petrifiés et metallisés
enfermés des pierres &c.

[57v] Ce ne sont pas des jeux de la nat.[:] prouvé par la texture[,] par la composn. chimique.
Ils ne forment pas eux mêmes les lits qui les contiennent.

[58r] Prop.
les parties qui ne contiennent point de corps organisés sont les plus anciennes.
donc l'organisation n'a pas toujours existé.

[59r] Prop.
les mont. granit. ne sont point par couches

[60r] Prop.
Il y a eu plusieurs changemens d'état successifs de la mer en terre[,] de terre en mer.
et dans une seule et même mer.

[61r] Prop.
Plusieurs des révolutions qui ont changé l'état du globe ont été subites.

[62r] Prop.
Il y a eu différens âges, produisans des genres de fossiles différens.

[63r] Prop.
Les fossiles ont souvent été déposés dans une eau tranquille; et n'ont point été transportés;

[64r] Prop.
les causes encore actuellement actives [ou] supposées telles, ne peuvent avoir produit les changemens dont on trouve des traces.
    ni les volcans
    ni le flux et le reflux
    ni la marche de la mer vers l'occident
    ni les inondations des fleuves &c.
    ni les alluvions
    ni les lithophytes
de la retraite de la mer.

[65r] *notes éparses*
on n'a encore creusé qu'à environ un 1/6000 du diamètre de la terre.

le falun de touraine s'étend sur 9 lieues carrées à 20 pieds de profondeur.

les couches de cailloux roulés[,] quelquefois redressés

on trouve des coquilles
    à l'etna, à 2400 pieds
    aux andes, à 14022 pieds
    aux pyrénées

[65v] la montagne de la table est granitique[;] la plupart des montagnes au nord également

les arbres enfouis dans les tourbières de Hollande et de Westphalie sont couchés du sud est au nord ouest.

SOURCES FOR FIGURES

1. Bultingaire, "Iconographie de Cuvier" (1932), pl. 1a.
2. Cuvier, *Briefen an Pfaff* (1845), p. 246. The wood engraving was presumably traced or adapted from a manuscript sketch in the original letter.
3. Bultingaire, "Iconographie de Cuvier" (1932), pl. 1b. Ardouin, *Georges Cuvier* (1970), suggests it is a self-portrait.
4. Cuvier, "Espèces d'éléphans" (1799), pl. IV.
5. Cuvier, "Espèces d'éléphans" (1799), pl. VI.
6. Cuvier, "Squelette trouvé au Paraguay" (1796). The engraving is smaller and cruder than Bru's original, which is reproduced in Rudwick, *Scenes from deep time* (1992), fig. 12.
7. Cuvier, "Squelette trouvé au Paraguay" (1796), pl. 2.
8. A loose sheet in Cuvier's file of notes for "Palaeothériums et Anoplothériums," MS 628, Bibliothèque Centrale, Muséum National d'Histoire Naturelle, Paris.
9. Cuvier, "Os fossiles dans la pierre à plâtre" (1804–8), vol. 3, pl. 46.
10. Cuvier, "Rétablissement des squelettes," in *Ossemens fossiles* (1812), vol. 3, 7th memoir, first (unnumbered) plate.
11. Cuvier's manuscript drawing, MS 635, Bibliothèque Centrale, Muséum National d'Histoire Naturelle, Paris.
12. Cuvier, "Petit quadrupède du genre de sarigues" (1804), pl. 19, figs. 4, 10.
13. Bultingaire, "Iconographie de Cuvier" (1932), pl. 2.
14. Cuvier, "Grande mastodonte" (1806), pl. 53.
15. Cuvier, *Rapport historique* (1810), title page.

16. Cuvier and Brongniart, "Géographie minéralogique des environs de Paris" (1811), pl. 2, fig. 1.

17. Cuvier and Brongniart, "Géographie minéralogique des environs de Paris" (1811), central portion of map.

18. Cuvier and Brongniart, "Géographie minéralogique des environs de Paris" (1811), pl. 2, left (southern) half of "Coupe no. 2 de Paris à Montmorency."

19. Cuvier and Brongniart, "Géographie minéralogique des environs de Paris" (1811), part of pl. 2.

20. Cuvier, "Os fossiles de ruminans" (1809), part of pl. 34.

21. Cuvier, *Ossemens fossiles* (1812), vol. 1, title page.

22. Cuvier, *Ossemens fossiles* (1812), first page of "Discours préliminaire."

23. Cuvier, "Mémoire sur l'ibis" (1804), reprinted in *Ossemens fossiles* (1812), vol. 1, pl. 52.

24. Cuvier, "Mémoire sur l'ibis" (1804), reprinted in *Ossemens fossiles* (1812), vol. 1, pl. 53.

25. Cuvier, *Discours sur les révolutions* (1826), frontispiece.

INDEX

Secondary formations, 79, 100, 105, 107,
118–19, 123–24, 158–59, 177, 188, 203–4,
223, 249, 251
sections, geological, 129–30, 138
Seine, 7, 133–34, 138
Sète, 56
Sèvres, 128, 135, 140, 144
shells, fossil. *See* mollusks
Siberia, 21, 51–53, 56, 89, 93, 119
Siliceous Limestone, 130–31, 153
Simplicius, 242
Sloan, Hans, 47
sloth, 26, 29–32, 53, 215
Smith, Jean, 270
Smith, William, 129n, 131n, 134n, 137n
Société d'Histoire Naturelle, 33
Société Philomathique, 33
Sömmerring, Samuel, 206n, 217n
Solnhofen. *See* Eichstatt
Somme, 56, 163
Soulavie, Jean Giraud-, 116
Spallanzani, Lazzaro, 52, 120, 232
species, creation of, 77, 229, 258; definition of,
226; origin of, 44, 57, 260–61
stalactite, 51–52, 159, 196, 225
Sternberg, count, 124
Strabo, 209, 235–36
stratigraphy, 4n
Stuttgart, 2
subordination of characters, 26, 36n
superficial deposits, 16, 44, 49, 52, 68, 90, 94,
159–61, 224, 235. *See also* alluvial deposits;
Detrital Silt
systems, geological. *See* theory of the earth

tapir, 32, 34, 39, 40–41, 55, 60, 66–67, 71,
90–91, 94, 96, 208, 215
Tartary, 244
Terror, period of. *See* Revolution, French
textual criticism, 81, 178, 215
theory of the earth, 6, 17, 61, 72, 82, 99, 104,
115, 126, 176–78, 256, 258, 265. *See also*
geology
Thuringia, 157, 251
Tibet, 93
Tidal waves, 202
tiger, 55, 159, 232

timescale, geological, 5, 6n, 53, 65, 68, 70, 77,
80, 105, 174, 179, 228–29, 259. *See also*
chronology, human; chronometers, natural
Tonna, 92
Touraine, 85
transformism, x–xi, 17, 48, 60, 76, 82–83, 91,
96, 168, 179–80, 226–31, 257–58, 261, 264
transportation of species, 48, 96
trigonia, 177n
tsunamis. *See* tidal waves
turtle, 54, 148, 157

unicorn, 213
United States, 21n, 26n, 51, 53
Urals, 93

Valais, 109
Valdarno, Society of, 171n
variation within species, 179, 226
veins, mineral, 118
Venice, 236
Verona, 56, 123, 251
Vesuvius, 120
Vicuña, 208, 215
Vincent, François, 75
Vivarais, 121
volcanos, 120, 197
Voltaire, 54n, 220n
Vosges, 109–10, 116, 250

Werner, Abraham, 4, 7, 114–15, 121, 128, 151,
158, 204, 223
Westphalia, 86
Whiston, William, 46, 125, 199n
Wiedemann, Christian, 10, 57
whale, 54, 56, 159
wolf, 54, 159, 226
wombat, 67, 208, 229
Woodward, John, 46, 125, 199n
Württemberg, 1–3

yak, 210

Zadig, 220
Zealand, 122
zebra, 210
zodiacs, 181, 246–47